2015-2016
ASSESSMENT OF THE ARMY RESEARCH LABORATORY

Army Research Laboratory Technical Assessment Board

Laboratory Assessments Board

Division on Engineering and Physical Sciences

A Report of

The National Academies of
SCIENCES · ENGINEERING · MEDICINE

THE NATIONAL ACADEMIES PRESS
Washington, DC
www.nap.edu

THE NATIONAL ACADEMIES PRESS **500 Fifth Street, NW** **Washington, DC 20001**

This activity was supported by Contract No. W911QX-14-C-0039 from the Army Research Laboratory. Any opinions, findings, conclusions, or recommendations expressed in this publication do not necessarily reflect the views of any organization or agency that provided support for the project.

International Standard Book Number-13: 978-0-309-45436-0
International Standard Book Number-10: 0-309-45436-0
Digital Object Identifier: https://doi.org.10.17226/24653

Additional copies of this publication are available for sale from the National Academies Press, 500 Fifth Street, NW, Keck 360, Washington, DC 20001; (800) 624-6242 or (202) 334-3313; http://www.nap.edu.

Printed in the United States of America

Suggested citation: National Academies of Sciences, Engineering, and Medicine. 2017. *2015-2016 Assessment of the Army Research Laboratory*. Washington, DC: The National Academies Press. https://doi.org.10.17226/24653.

The National Academies of
SCIENCES • ENGINEERING • MEDICINE

Reports document the evidence-based consensus of an authoring committee of experts. Reports typically include findings, conclusions, and recommendations based on information gathered by the committee and committee deliberations. Reports are peer reviewed and are approved by the National Academies of Sciences, Engineering, and Medicine.

Proceedings chronicle the presentations and discussions at a workshop, symposium, or other convening event. The statements and opinions contained in proceedings are those of the participants and have not been endorsed by other participants, the planning committee, or the National Academies of Sciences, Engineering, and Medicine.

For information about other products and activities of the National Academies, please visit nationalacademies.org/whatwedo.

ARMY RESEARCH LABORATORY TECHNICAL ASSESSMENT BOARD

JENNIE S. HWANG, NAE,[1] H-Technologies Group, *Chair*
KENNETH R. BOFF, Socio-Technical Sciences
MARK EBERHART, Colorado School of Mines
GEORGE (RUSTY) T. GRAY III, Los Alamos National Laboratory
PRABHAT HAJELA, Rensselaer Polytechnic Institute
WESLEY L. HARRIS, NAE, Massachusetts Institute of Technology
ALAN NEEDLEMAN, NAE, Texas A&M University

Staff

LIZA HAMILTON, Associate Program Officer
EVA LABRE, Administrative Coordinator
JAMES P. McGEE, Director
ARUL MOZHI, Senior Program Officer

PANEL ON BALLISTICS SCIENCE AND ENGINEERING AT THE ARMY RESEARCH LABORATORY

GEORGE (RUSTY) T. GRAY III, Los Alamos National Laboratory, *Chair*
CHARLES E. ANDERSON, JR., CEA Consulting
MELVIN R. BAER, Sandia National Laboratories (retired)
CAMERON R. (DALE) BASS, Duke University
ELLIOT L. CHAIKOF, NAM,[2] Harvard Medical School
J. GARY EDEN, NAE, University of Illinois, Urbana-Champaign
ROGER GHANEM, University of Southern California
ZELDA GILLS, Lockheed Martin Aeronautics Company
MICHAEL JAFFE, New Jersey Institute of Technology
JERRY A. KRILL, Johns Hopkins University, Applied Physics Laboratory
WILLIAM S. MARRAS, NAE, Ohio State University
ALLAN T. MENSE, Raytheon Missile Systems
GARY L. MESSING, Pennsylvania State University
ALAN NEEDLEMAN, NAE, Texas A&M University
DONALD PROSNITZ, Independent Consultant
GEORGE C. SCHATZ, NAS,[3] Northwestern University
FRANK J. SERNA, Charles Stark Draper Laboratory
JOSEPH E. SHEPHERD, California Institute of Technology
LESLIE E. SMITH, National Institute of Standards and Technology (retired)
STEVEN F. SON, Purdue University
GHATU SUBHASH, University of Florida
TRACY J. VOGLER, Sandia National Laboratories
ELIHU ZIMET, Independent Consultant

[1] Member, National Academy of Engineering.
[2] Member, National Academy of Medicine.
[3] Member, National Academy of Sciences.

PANEL ON HUMAN FACTORS SCIENCE AT THE ARMY RESEARCH LABORATORY

KENNETH R. BOFF, Socio-Technical Sciences, *Chair*
HERBERT H. BELL, Air Force Research Laboratory (retired)
BRUCE G. BERG, University of California, Irvine
MICHAEL L. BONINGER, NAM, University of Pittsburgh Medical Center
LINDA NG BOYLE, University of Washington
BARRETT S. CALDWELL, Purdue University
BARBARA A. DOSHER, NAS, University of California, Irvine
FRANCIS (FRANK) T. DURSO, Georgia Institute of Technology
HOWARD EGETH, Johns Hopkins University
RALPH R. ETIENNE-CUMMINGS, Johns Hopkins University
JOHN H. (JACK) HAMMOND, Lockheed Martin (retired)
MOSHE KAM, New Jersey Institute of Technology
JOHN LACH, University of Virginia
MARGARET L. LOPER, Georgia Tech Research Institute
WILLIAM W. LYTTON, SUNY Downstate Medical Center
STEPHEN C. MERRIMAN, The Boeing Company
KATHERINE L. MORSE, Johns Hopkins University, Applied Physics Laboratory
SHRIKANTH (SHRI) S. NARAYANAN, University of Southern California
BARBARA G. SHINN-CUNNINGHAM, Boston University
MARY C. WHITTON, University of North Carolina, Chapel Hill
KING-HAY YANG, Wayne State University

PANEL ON INFORMATION SCIENCE AT THE ARMY RESEARCH LABORATORY

PRABHAT HAJELA, Rensselaer Polytechnic Institute, *Chair*
TILAK AGERWALA, International Business Machines Corporation (retired)
JEAN M. ANDINO, Arizona State University
DAVID W. AUCSMITH, University of Washington
PETER A. BELING, University of Virginia
RONALD J. BRACHMAN, Cornell Tech
KATHLEEN M. CARLEY, Carnegie Mellon University
JOHN M. CARROLL, Pennsylvania State University
VINTON G. CERF, Google, Inc.
BRONIS R. DE SUPINSKI, Lawrence Livermore National Laboratory
MARK E. DEAN, NAE, University of Tennessee, Knoxville
MICHAEL J. FRANKLIN, University of California, Berkeley
IRENE GREIF, NAE, IBM Thomas J. Watson Research Center (retired)
JONATHAN GRUDIN, Microsoft Research
VERLIN B. HINSZ, North Dakota State University
BARRY M. HOROWITZ, NAE, University of Virginia
MARY JANE IRWIN, NAE, Pennsylvania State University
JAMES T. KAJIYA, NAE, Tolt Machine Works, Inc.

GEORGE KARYPIS, University of Minnesota
CHARLES E. KOLB, NAE, Aerodyne Research, Inc.
VIPIN KUMAR, University of Minnesota
JOHN M. LANICCI, Embry-Riddle Aeronautical University
GUY M. LOHMAN, IBM Research—Almaden Center
KYRAN D. MISH, Sandia National Laboratories
J. GREGORY MORRISETT, Cornell University
RANDOLPH L. MOSES, Ohio State University
JOSE M.F. MOURA, NAE, Carnegie Mellon University
LINDA A. NESS, Independent Consultant
AROGYASWAMI J. PAULRAJ, NAE, Stanford University
RADIA PERLMAN, NAE, EMC Corporation
KALYAN PERUMALLA, Oak Ridge National Laboratory
STEVEN J. WALLACH, NAE, Micron Technology, Inc.

PANEL ON MATERIALS SCIENCE AND ENGINEERING AT THE ARMY RESEARCH LABORATORY

MARK E. EBERHART, Colorado School of Mines, *Chair*
RICHARD C. ALKIRE, NAE, University of Illinois, Urbana-Champaign
STANLEY ATCITTY, Sandia National Laboratories
MICHAEL BASKES, NAE, Mississippi State University
JOSEPH J. BEAMAN, JR., NAE, University of Texas, Austin
WILLIAM J. BOETTINGER, NAE, National Institute of Standards and Technology (retired)
PAUL V. BRAUN, University of Illinois, Urbana-Champaign
LALIT C. CHHABILDAS, Air Force Research Laboratory (retired)
DIANNE CHONG, The Boeing Company (retired)
STEPHEN CREAGER, Clemson University
NEIL A. DUFFIE, University of Wisconsin, Madison
HORACIO D. ESPINOSA, Northwestern University
MICHAEL ETTENBERG, NAE, Dolce Technologies
JAMES S. HARRIS, NAE, Stanford University
W. JOHN KAO, University of Wisconsin, Madison
THOMAS L. KOCH, NAE, University of Arizona
CHRISTIAN MAILHIOT, Sandia National Laboratories
CHRIS A. MARIANETTI, Columbia University
MARC A. MEYERS, University of California, San Diego
DAVID J. MOONEY, NAE/NAM, Harvard University
DUNCAN T. MOORE, NAE, University of Rochester
KARIN M. RABE, NAS, Rutgers, State University of New Jersey
KRISHNA RAJAN, University of Buffalo, State University of New York
ALAN S. RUDOLPH, Colorado State University
ROBERT E. SCHAFRIK, NAE, General Electric Aviation (retired)
STEWART A. SILLING, Sandia National Laboratories
NARESH N. THADHANI, Georgia Institute of Technology

PANEL ON MECHANICAL SCIENCE AND ENGINEERING AT THE ARMY RESEARCH LABORATORY

WESLEY L. HARRIS, NAE, Massachusetts Institute of Technology, *Chair*
RUZENA K. BAJCSY, NAE/NAM, University of California, Berkeley
L. CATHERINE BRINSON, Northwestern University
FRANCESCO BULLO, University of California, Santa Barbara
FU-KUO CHANG, Stanford University
GLENN S. DAEHN, Ohio State University
GREGORY D. HAGER, Johns Hopkins University
RAINER HEBERT, University of Connecticut
AYANNA M. HOWARD, Georgia Institute of Technology
AGNES C. KLUCHA, UTC Aerospace Systems
WILLIAM Q. MEEKER, JR., Iowa State University
DAVID A. PETERS, Washington University
ARMANDO A. RODRIGUEZ, Arizona State University
KENNETH M. ROSEN, NAE, Aero-Science Technology Associates, LLC
JAYANT S. SABNIS, Massachusetts Institute of Technology
MICHAEL SALKIND, Indus International, Inc.
F. STAN SETTLES, NAE, University of Southern California
JULIE A. SHAH, Massachusetts Institute of Technology
WILLIAM A. SIRIGNANO, NAE, University of California, Irvine
EDWARD C. SMITH, Pennsylvania State University
CHARLES E. THORPE, Clarkson University
RYAN B. WICKER, University of Texas, El Paso
DENNIS F. WILKIE, NAE, Ford Motor Company and Motorola, Inc. (retired)
HOLLY YANCO, University of Massachusetts, Lowell
BEN T. ZINN, NAE, Georgia Institute of Technology

PANEL ON ASSESSMENT AND ANALYSIS AT THE ARMY RESEARCH LABORATORY

ALAN NEEDLEMAN, NAE, Texas A&M University, *Chair*
STEVEN M. BELLOVIN, NAE, Columbia University
CARY F. CHABALOWSKI, Michigan Technological University
LESTER A. FOSTER III, Electronic Warfare Associates
ROGER GHANEM, University of Southern California
TERRY P. LEWIS, Raytheon Company
STEVEN B. LIPNER, Independent Consultant
GREG H. PARLIER, Independent Consultant
JILL H. SMITH, U.S. Army Communications-Electronics RDE Center (retired)
JAMES J. STREILEIN, Independent Consultant

LABORATORY ASSESSMENTS BOARD

JOHN W. LYONS, NAE, National Defense University (retired), *Chair*
ROSS B. COROTIS, NAE, University of Colorado, Boulder
PAUL A. FLEURY, NAS/NAE, Yale University
C. WILLIAM GEAR, NAE, Princeton University
WESLEY L. HARRIS, NAE, Massachusetts Institute of Technology
JENNIE S. HWANG, NAE, H-Technologies Group
W. CARL LINEBERGER, NAS, University of Colorado, Boulder
C. KUMAR N. PATEL, NAS/NAE, Pranalytica, Inc.
ELSA REICHMANIS, NAE, Georgia Institute of Technology
LYLE H. SCHWARTZ, NAE, Air Force Office of Scientific Research (retired)

Staff

LIZA HAMILTON, Associate Program Officer
EVA LABRE, Administrative Coordinator
JAMES P. McGEE, Director
ARUL MOZHI, Senior Program Officer

Acknowledgment of Reviewers

This report has been reviewed in draft form by individuals chosen for their diverse perspectives and technical expertise. The purpose of this independent review is to provide candid and critical comments that will assist the institution in making its published report as sound as possible and to ensure that the report meets institutional standards for objectivity, evidence, and responsiveness to the study charge. The review comments and draft manuscript remain confidential to protect the integrity of the deliberative process. We wish to thank the following individuals for their review of this report:

Rodney J. Clifton, NAE, Brown University,
Nancy J. Cooke, Arizona State University,
Earl H. Dowell, NAE,[1] Duke University,
William D. Gropp, NAE, University of Illinois, Urbana-Champaign,
W. Carl Lineberger, NAS,[2] University of Colorado, Boulder,
Dwight C. Streit, NAE, University of California, Los Angeles, and
Donna K. Vargas, Las Cruces, New Mexico.

Although the reviewers listed above have provided many constructive comments and suggestions, they were not asked to endorse the conclusions or recommendations, nor did they see the final draft of the report before its release. The review of this report was overseen by David E. Crow, NAE, University of Connecticut, who was responsible for making certain that an independent examination of this report was carried out in accordance with institutional procedures and that all review comments were carefully considered. Responsibility for the final content of this report rests entirely with the authoring board and the institution.

---

[1] Member, National Academy of Engineering.
[2] Member, National Academy of Sciences.

Contents

APPENDIXES

Summary

The statement of task that guided the work of the Army Research Laboratory Technical Assessment Board (ARLTAB) is as follows:

> An ad hoc committee to be named the Army Research Laboratory Technical Assessment Board (ARLTAB), to be overseen by the Laboratory Assessments Board, will be appointed to continue the function of providing biennial assessments of the scientific and technical quality of the Army Research Laboratory (ARL). These assessments will include findings and recommendations related to the quality of ARL's research, development, and analysis programs. While the primary role of the ARLTAB is to provide peer assessment, it may offer advice on related matters when requested by the ARL Director. The ARLTAB will provide an interim assessment report at the end of Year 1 of each 2-year assessment cycle and a final assessment report biennially. The ARLTAB will be assisted by up to seven separately appointed panels that will focus on particular portions of the ARL program. Each year, up to three additional panels may be appointed to assess special topics, at the request of the ARL Director.

During the 2015-2016 assessment, the ARLTAB was assisted by six panels, each of which focused on a portion of the ARL program conducted in ARL's science and technology (S&T) campaigns: Materials Research, Sciences for Lethality and Protection, Information Sciences and Computational Sciences, Sciences for Maneuver, Human Sciences, and Analysis and Assessment. This report summarizes the findings of the Board for the 2015-2016 biennial assessment.

The mission of ARL, as the U.S. Army's corporate laboratory, is to provide innovative science, technology, and analyses to enable a full spectrum of operations. In 2013, ARL restructured its portfolio of ongoing and planned research and development (R&D) to align with its S&T campaign plans for 2015-2035. ARL has maintained its organizational structure, consisting of six directorates: Weapons and Materials Research Directorate (WMRD), Computational and Information Sciences Directorate (CISD), Human Research and Engineering Directorate (HRED), Sensors and Electron Devices Directorate (SEDD), Survivability and Lethality Analysis Directorate (SLAD), and Vehicle

Technology Directorate (VTD). The research portfolio has been organized into science and technology campaigns, each of which describes related work supported by staff from multiple directorates. Appendix Table A.1 shows the directorates that supported each campaign during the 2015-2016 review. ARL's technical strategy document describes the portfolio of each campaign in detail.[1] ARL's vision is compelling and raises expectations for an innovative program of research designed to be responsive to the needs of the Army after next. This is not yet fully evident in the portfolio currently being assessed. The reorganization of the portfolio into key focused campaigns is promising, but it may take some time to transform and mature the program of work to consistently align with new critical paths.

In general, the quality of the research presented, the capabilities of the leadership, the knowledge and abilities of the investigators, and proposed future directions continue to improve. Significant gains were evident in publication rates, numbers of postdoctoral researchers, and collaborations with relevant peers outside ARL. The research work environments were impressive in terms of their unique and advanced technology capabilities to support research. Overall these are all outstanding accomplishments and mark an advance over prior years.

MATERIALS RESEARCH

ARL's materials research efforts span the spectrum of technology maturity as they address Army applications, working from the state of the art to the art of the possible—25 years out, according to ARL. Materials research efforts and expertise, one of ARL's core technical competencies, are appropriately spread throughout the ARL enterprise.

Biological and Bioinspired Materials

The biological and bioinspired materials effort, though small, has an excellent track record, including the stabilization of proteins against thermal and chemical extremes using new chemistries and methods to derive antibody-like reagents that improve antibody properties—accomplishments that are likely to lead to program growth. The scientific quality of this thrust area is on par with the work at leading federal, university, and industry laboratories and is a crucial part of a broader national effort in biomaterials research. Because biology is a growth area, ARL now has an opportunity to identify and recruit a critical mass of biologists, including microbiologists and polymer/organic chemists, looking well into the future to create an integrated community of researchers.

Energy and Power Materials

Energy and power materials is a broad mission-critical research area covering a range of different devices, fuels, and applications across a wide range of time and length scales. The research portfolio supports an appropriate balance of high-risk, long-term-impact projects along with mid- and short-term projects. The quality of the research is comparable to that of top academic and industrial research laboratories. However, there are questions as to whether ARL is mobilizing aggressively enough to capitalize on ARL's internal advances and as well as external advances made by the broader community—for example, in quantum-well infrared photo detectors and in silicon photonics. In general, the programs are

[1] U.S. Army Research Laboratory, *Army Research Laboratory Technical Strategy 2015-2035*, Adelphi, Md., 2014, http://www.arl.army.mil/www/pages/172/docs/ARL_ Technical Strategy_FINAL.pdf.

working to integrate modeling with experimental study, but in contrast to the expansion in first-principles modeling, engineering models are underutilized.

Engineered Photonics Materials

ARL's engineered photonics materials research effort is among the world's best. This is an impressive accomplishment in light of the technical program's inherently wide scope. The quality of the work presented reflects a high level of technical competence and professionalism on the part of researchers and management and shows a good balance of high-risk, longer-term work and nearer-term, customer-driven solutions and incremental—but critical—technology refinement. This well-balanced portfolio is supported by a strong materials capability in terms of both staff expertise and facilities. Investments in computational modeling and simulation have been successfully implemented to complement strengths and core competencies in materials synthesis and characterization, as well as device work. All of these facilities and capabilities are being leveraged into compelling device- and application-driven work, especially in ultraviolet materials and infrared devices and device physics in both areas.

High Strain Rate and Ballistic Materials

ARL's high strain rate (HSR) and ballistic materials effort is excellent, having established itself as a world leader by building novel capabilities, including, for example, extensive facilities for metals, polymers, and composites processing. The miniaturized Hopkinson bar and multiscale rate-dependent mechanical testing equipment—along with microscale sample preparation set-up for investigating polymers, metals, ceramics, fibers, and threads—are unique facilities. The commitment by ARL to take advantage of the dynamic sector facilities at the Advanced Photon Source is noteworthy. In situ measurements performed using these facilities will provide the needed fundamental knowledge for developing and validating computational models for improved understanding of HSR effects. Throughout the HSR and ballistic materials efforts, there is substantial growth in the use of computation and modeling and its integration with experimentation. Continued advances in this area are needed, with the addition, wherever possible, of physics-based analysis.

Structural Materials

The research of the structural materials effort is successfully coupling modeling with experimentation. Of note are programs designed not only to support specific and narrowly focused materials development efforts but those that will produce more broadly applicable tools—for example, the program directed toward grain boundary modeling of ceramics for light-weight protective materials. The computational tools developed as part of this effort will be applicable to the study of grain boundary interfacial relationships across all ceramic materials. The potentially wide applicability for these modeling and simulation techniques will provide ARL with the capabilities it needs to respond rapidly to future threats.

Electronic Materials

Overall, the quality of applied R&D efforts in electronic materials is outstanding with well-supported staff. Long-range projects that are maturing and moving into manufacturing (MANTECH programs) are balanced by new advanced research initiatives.

The R&D under this thrust is directed, in part, to responding to dramatic changes in the battlefield environment and has transformed from a relatively small number of large, well-supported divisions to smaller distributed groups. This transformation challenges the Army to provide the infrastructure for supplies, particularly equipment repair. An exciting example of one of the research programs intended to meet this need is the application of basic metallurgical science to provide field deployable, individual custom repairs for damaged vehicles, including helicopters and armored vehicles.

SCIENCES FOR LETHALITY AND PROTECTION

The ARL research efforts in sciences for lethality and protection that were assessed in 2015 and 2016 span basic research that improves our fundamental understanding of scientific phenomena and generates technology that supports battlefield injury mechanisms in the following areas: human response to threats and human protective equipment; directed energy programs; ballistics and blast programs that address weapon–target interactions and armor and adaptive protection developments; disruptive energetics and propulsion technologies; effects on target—ballistics and blast; and flight, guidance, navigation, and control.

Battlefield Injury Mechanisms

A better understanding of battlefield injury mechanisms is vital to improving protective measures, making this program an important one for ARL. This is especially true for protection of the head, about which there is considerable uncertainty concerning allowable levels of shock, which greatly affect the protective options. The most impressive accomplishment of the program for battlefield injury mechanisms, human response, and human protective equipment is that a strong cadre of scientists is at work, and a credible program is under way. A long-term vision for the battlefield injury mechanisms projects could better guide resource allocation and program direction.

Directed Energy

ARL's campaign plans categorize directed energy (DE) as a focused area under the much broader category of electronic warfare (EW), in accordance with the Army's definitions. The ARL posture designations for both radio frequency-DE and laser-DE are *collaborate* rather than *lead*. ARL needs to take a strategic look at the area of DE to determine its priority going forward and rethink its effort with a view to the 2035 time frame. The strategic review needs to consider future capabilities that the Army will need that DE might enable, as well as what DE capabilities might be fielded by our adversaries for which the Army will need countermeasures. A focused ARL DE program would benefit from a systems-level study addressing future Army missions in which DE could play a role and where DE effectiveness and alternatives to DE are traded off.

Armor and Adaptive Protection

ARL has a strong record of achievement in the basic and applied sciences and the engineering of penetration and protection. The R&D described in the armor and adaptive protection area showed how ARL is building on its tradition of excellence to provide the knowledge basis for current and future Army needs in protecting our warfighters. This remains a core competency that underlies Army capabilities across the entire Department of Defense (DOD), and it needs to be preserved and nurtured. There was significant evidence of teamwork and integration among the projects in, for example, adaptive protection. Examples of the linkage between experimentation and computational modeling to provide physical insight into problems were especially noteworthy and had the potential to aid in developing new designs and exploring new concepts. Developing a predictive capability for damage and fracture in metals, ceramics, and polymers underlies the efficient development of new material systems for protection from emerging penetration capabilities.

Disruptive Energetics and Propulsion Technologies

ARL's synthesis effort is a commendable, relatively new effort at ARL to develop a chemical synthesis effort whose growth is warranted. A blended focus on various applications (propellants and explosives) is needed rather than just explosives. This is a high-risk/high-payoff effort, so ARL could expect that most candidate materials may not, ultimately, transition to Army applications and systems. The propellant simulation R&D remains a traditional strength of ARL that is positively impacting and supporting warfighter and Army needs, and it is necessary for it to be supported and nurtured. In the extended solids focus area, it is commendable that ARL has scaled some materials to more significant quantities (grams), and ARL needs to pursue scaling to larger quantities for testing where possible.

ARL could beneficially complement the experimental efforts to date in its energetics and propellant projects with modeling efforts that might suggest alternate geometries (e.g., cylindrical) and perhaps allow additional information (e.g., model parameters) to be obtained from the data. Developing 3D additive manufacturing techniques for solid propellants and temperature-sensitive glues is exciting and necessary.

Flight, Guidance, Navigation, and Control

ARL's research team has made significant progress toward developing the technical underpinnings of advanced guided munitions in the areas of aerodynamics, guidance and control, and terminal homing. ARL has attracted some outstanding personnel in the Flight, Guidance, Navigation, and Control (GN&C) area, especially new Ph.D.'s, and they are undertaking interesting and relevant work on par with academic departments. ARL needs to continue to invest in its staffing in this manner. The research in the areas of vortex fin interactions and highly maneuverable, small-diameter munitions is impressive. In the GN&C R&D area, ARL needs to decide in which areas it will lead, versus collaborate, versus watch or follow developments. Strategic thinking by ARL in the GN&C area is critically important to define the state of the art in aspects of GN&C that are important to the Army and determine where ARL will engage and at what level and with whom.

INFORMATION SCIENCES

The research portfolio in information sciences includes research projects in broad categories of system intelligence and intelligent systems (SIIS), sensing and effecting, network and communications (NC), human and information interaction (HII), cybersecurity, and atmospheric sciences.

System Intelligence and Intelligent Systems

Research in information understanding and retrieval is focused on methods for extracting temporal relationships from text for constructing knowledge networks, and using agent-based semantic analysis for information retrieval. Other projects are directed at identifying mechanisms for trust formation in human networks. Work on information fusion seeks to combine data from disparate sources and uses an approach of estimating credibility through fusion of subject opinions. Research is also directed at approaches for reasoning in an uncertain environment. Work on designing optimal paths for autonomous mobile robot movement is of very high quality.

Sensing and Effecting

In sensing and effecting, research in the application of new materials and deployment of innovative new signal-processing techniques has enhanced the effectiveness of acoustic sensors. The work on cross-modal face recognition is of high quality and relevant to Army missions. Development of new sensor algorithms for polarization imagery and manmade object discrimination has important practical implications. In radar signal processing, the use of nonlinear harmonic radar to achieve greater sensitivity across a narrow frequency band has shown promise. Similarly, work related to fusing data from multiple sources to improve inference has good potential.

Networks and Communications

Research projects in the NC area focus on understanding and exploiting the interactions between information and socio-technical networks, in particular, on communications and command and control networks. In research related to channels and protocols, there is a focus on very-high frequency (VHF) range communications, ultraviolet (UV) communications, and on quantum methods for networked communications. Work on UV communications includes an experimental program to validate theoretical propagation models. Research in network-based information processing has a focus on emulation and simulation tools for experimentation, information theoretic foundations, trust in networks, and decentralized learning. The Network Science Research Laboratory is a significant resource for this research.

Human and Information Interaction

HII research is a new endeavor within ARL and does not have fully developed areas and lines of research. The ongoing research projects represent the following three major focus areas: (1) the use of computational technologies to facilitate communications between humans and devices in a degraded information environment using natural language, and using images as appropriate; (2) the decision-making process in an environment where information is received from the built environment, social media, traditional media, and a myriad of sensors; and (3) an examination of how groups shape the information environment, gain information dominance, take actions, and affect appropriate response.

Cybersecurity

Research in cybersecurity is focused around theoretical advances and model development related to cyber threat detection, recognition, and defeat mechanisms. It examines detection and defeat of sophisticated attacks that are different from those encountered in commercial or civilian settings. The combination of strategic and tactical networks in use by the Army creates unique challenges in the domain of cybersecurity, including a constantly evolving environment of threats. The ARL portfolio of research in this area contains an even balance between theoretical and applied components.

Atmospheric Sciences

The research in atmospheric sciences has a focus on both the characterization and prediction of the diverse battlespace environment. Individual projects address the challenges of collection and processing of environmental data from nontraditional observing platforms, imaging and sensing of aerosols and objects in the battlespace environment, and understanding complex atmospheric flows using a combination of field observation experiments and model development. The emphasis is on developing accurate, relevant, and timely predictions on spatial and temporal scales useful to Army operations. The research portfolio includes a mix of analytical and computational projects as well as experimental projects.

COMPUTATIONAL SCIENCES

In the research portfolio under computational sciences, projects in the broad categories of advanced computing architectures, computing sciences, data-intensive sciences, and predictive simulation sciences were reviewed.

Advanced Computing Architectures

In the area of advanced computing architectures, ARL research has focused on tactical high-performance computing (HPC) and on exploring new computer architectures, including neurosynaptic computing, the epiphany of a many-core chip, and quantum networking. Research on computation ferrying is directed at tactical HPC—the issue of computational tasks that could be computed on handheld devices or on mobile tactical HPC resources. Beyond establishing modeling and simulation capabilities to guide offload decisions, critical issues related to programmability and performance of edge (handheld) devices are being explored. Research in dynamic binary translation is focused on allowing fast cross-architecture execution of binary codes, and it has yielded dramatic improvement in performance.

Computing Sciences

The computing sciences group has established a strategic focus in quantum computing, parallel processing environments for large heterogeneous parallel systems, and tools to simplify application development for HPC environments. Research on the development of a threaded message-passing interface for reduced instruction set computing array multicore processors has yielded innovative solutions to the challenge of power-efficient parallel programming. The work on HPC-scaled quantum hardware description language is representative of one of the few efforts in the area of quantum networking.

Data-Intensive Sciences

Accomplishments in data-intensive sciences include new model order reduction methods for partial differential equations (PDE), cognitively steered exploration of solutions to PDEs at different resolution, efficient summarization and visualization of high-dimensional data sets, and concise characterization of spall damage in materials. The work on developing a high-performance, sparse, nonnegative, least-square solver advances the state of the art and leverages ARL's expertise in numerical analysis and HPC. Similarly, the work on neuromorphic computing represents a fresh and original approach. Good advances have been made in large, multiscale material modeling, particularly to identify damage modes. The visual simulation laboratory focuses the use of a visualization-based framework to allow users to steer a multiresolution PDE simulation.

Predictive Simulation Sciences

Important contributions have been made to developing predictive capabilities for use on the low-power computer platforms available in the field. In materials modeling, research on scalable algorithms for simulating dislocations in microstructured crystals is promising and has broad applicability for material and structural failure simulations. One of the most difficult technical challenges for this work is the problem of effective load balancing of HPC resources, and the research could benefit from consideration of new developments in adaptive parallel load-balancing techniques.

SCIENCES FOR MANEUVER

Vehicle Intelligence

In each of the three pillars of the vehicle intelligence (VI) program—human–robot interaction, intelligence and control (I&C), and perception—the research quality was generally high. Collaboration with other government agencies, industry, and universities continues to have positive benefits. Internal personnel advancement strengthens the science capability for the VI R&D program. Each of the three pillars of the VI program has demonstrated significant progress in advancing its R&D objectives to support the warfighter in increasingly complex environments. The R&D activities are consistent with their defined objectives. Opportunities in multiperson/multirobot scenario simulation, the teaming of autonomous systems with soldiers in uncertain environments, multispectral sensing, range sensing, contact sensing, and immersive display of robot LIDAR imagery are likely to allow ARL to be of even greater benefit to the soldier.

Inclusion of soldiers in VI field experiments is commendable. Use of more realistic vignettes and real-life simulations in experiments would be very beneficial. In particular, the use of realistic war fighting vignettes, where researchers are in the field with soldiers, provides opportunities to test and evaluate research hypotheses more thoroughly, including the revelation of previous unknowns.

Some strategic goals and tactical milestones for VI R&D programs could be made more apparent. To help quantify general progress and application-specific performance, more efforts need to be made in baselining and benchmarking.

The Board's conclusion on I&C research at ARL is that the whole is considerably less than the sum of the parts. Challenges to I&C research focus on determining how to deal with trade-offs in order to determine which research to continue; how to effectively integrate outcomes from the individual projects and develop a methodology for this integration; how to share the overarching systems perspective and relay that vision to the research projects; how to identify and validate the process of getting from high-level capability or needs to research tasks (and evaluation or benchmarking of whether they comply with needs); how to appropriately delineate between basic and early applied work; how to balance and integrate top-down and bottom-up-driven processes; how to compare the research against the standard baseline data sets (when available) and identify standard metrics for validating whether the research has achieved the stated goals of the proposed work; and how to transition the research from work on simplified problems that facilitate analysis to actual scenarios that are germane to the Army's unique problems and characteristics.

Vehicle Technology

Similarly, in each of the foundational pillars and key enablers of the vehicle technology program—platform mechanics, energy and propulsion, platform mechanics, and logistics and sustainability—the research quality was generally high. Research results are published in high-quality journals. Collaboration with other government agencies, industry, and universities continues to yield positive benefit. Internal personnel advancement, including hiring new, well-qualified Ph.D. researchers, strengthens the capability of the Sciences for Maneuver R&D program.

The Koopman decomposition of periodically excited Hopf bifurcation research (nonlinear system theory) initiative is outstanding with potential significant impact on Army understanding and exploitation of nonlinear mechanics such as dynamic stall, Floquet instabilities, low-order modeling of control system design, and ground resonance instabilities.

The overall quality of the transient thermal management of electronic components is excellent. This work is critical to war-fighting missions. Using multilevel encapsulated phase change material, this work focuses on examining fast transients for high-power applications. A big picture demonstration showing how the energy and power work will be integrated would be helpful for future ARLTAB assessments and to researchers. Such a demonstration can be used to pursue a very relevant and informative system-of-systems analysis that can be used to identify weak links, guide research, and steer collaborations.

HUMAN SCIENCES

Six elements of human sciences were assessed during 2015-2016. In addition, a component of the Analysis and Assessment Campaign portfolio on assessing mission capabilities of systems was reviewed during 2015.

The overall technical quality of the Human Sciences Campaign is good and has shown continual improvement. ARL continues on a trajectory of hiring highly skilled postdoctoral researchers, many of whom are being groomed to become full-time ARL employees. Publication in peer-reviewed journals and participation at professional conferences has continued to grow, coupled with increasing participation in professional activities (e.g., journal editing). Collaborations with peer communities appear to be healthy and provide ARL personnel with invaluable networking opportunities and the options to leverage quality research elsewhere. ARL's investment in quality R&D in the human sciences has increased its potential for impact on the present and future Army.

Humans in Multiagent Systems

As scoped, the area of humans in multiagent systems is very broad. It includes interactions between humans and technology and between humans and other human beings (sociocultural interactions), and it is not clear how those pieces fit together. The challenges for human sciences R&D are in understanding and dealing with human factors such as soldier workload, situation awareness, trust, influence, and cultural cognition. This is an important interdisciplinary area with unquestionable Army relevance where ARL human sciences needs to have a lead role. A coherent vision of ARL's niches for ongoing and future research needs to be established.

Real-World Behavior

The collection and analysis of human behavioral data in dynamic, complex, natural environments is an ambitious and challenging undertaking. Not surprisingly, the accomplishments in this area are incremental given the immature state of the art and the challenges to developing the needed enabling technology and methodology. The research presented is well focused on mission-relevant problems and contexts and draws on measures from multiple domains (e.g., biomechanics, cognition, and neurosciences), which is consistent with the goal of addressing real-world complexity. Continued strategic investment in this area can yield significant payoffs for the Army with potential spillover benefits to other government and private sector R&D.

Toward Human Variability

Understanding and predicting human variability is an important and timely topic for investigation. Current systems are calibrated to the average performance of the average person in challenging circumstances; optimized adaptive systems might enable better use of human capacity when situations and states permit. Advances in this area reflect the availability of increasingly sophisticated techniques for behavioral and brain measurement and the development of new analytic and statistical methods that could enable adaptive systems in operational settings. This group has recruited an exceptionally strong set of researchers, including well-qualified postdoctoral and early-career scientists representing different technical backgrounds. Overall, the work in this area was of exceptional quality.

Training

The goal of the training program is to discover and develop methods, models, tools, and technologies that will increase soldier readiness by improving training methods and training technologies. The scope and complexity of training science and technology is massive.

The technology baseline on which training depends cuts across multiple science and technology areas and, in turn, requires a diverse, multidisciplinary workforce. Likewise, the relevant basic cognitive and social science research that applies to many training problems is fragmented and somewhat characterized by competing micro-theories. Overall the area is well staffed and professionally connected and is drawing upon appropriate methods and using relevant environments and subject populations. Generally, the training program is addressing selected scientific and technical challenges relevant to both the Army and the education and training community at large. In particular, the intelligent tutoring work stands out for its technical leadership and achievements.

Integration Technologies

The objectives of the integration technology R&D program are to discover and innovate principles and mechanisms for integrating humans and systems. The area has assembled a highly qualified and productive team with expertise in electroencephalogram, statistical and neural network modeling, and the modeling of neural substrates of behavior. It has employed quality research methods in physiological and behavioral signal measurement and analyses and in advanced machine-learning techniques. Competitive advances in the characterization and understanding of the human state (cognitive, affective, and physical) from neural and behavioral measurements are significant, its publications are strong, and the potential return on investment in this area is very high.

Augmentation

The goals of the augmentation R&D program are to develop and enable technological approaches to augmenting fundamental human capabilities that may enhance Army mission-related performance. The scientists and engineers working in this area have demonstrated keen awareness of the trade-offs between the burden of additional gear, machinery, weight, and maintenance against the expected benefits from augmentation devices and algorithms. The ARL augmentation team is well positioned to become a leading global force in the R&D of augmentation for healthy individuals. These enhanced performance capabilities offer significant potential benefits to the Army and opportunities for valuable downstream spin-offs to the civilian sector.

Assessing Mission Capabilities of Systems

In the area of assessing mission capabilities of systems, the work comprises human-centered engineering and decision support methods, models, and tools supported under ARL's Analysis and Assessment Campaign. Most of the projects that were presented in this area are responding to the needs of specific Army customers. Commendable efforts are under way at ARL to advance assessment science by developing new models, tools, and metrics to support the acquisition and fielding of effective human–machine systems responsive to emerging missions and threats. ARL has the opportunity to be on the forefront of the research in this area; however, the current portfolio of projects in human–system integration (HSI) may be too customer-driven. ARL could leverage this applied work and/or fund companion projects to advance the state of scientific knowledge for HSI and to broaden the impact of the work beyond the immediate customers.

ANALYSIS AND ASSESSMENT

ARL's Analysis and Assessment Campaign provides tools that increase awareness of material capabilities, assess the survivability and lethality of Army systems, and improve and simplify the Army's decision making. Generally, ARL needs to broaden the perspectives of the campaign staff members. ARL also needs to acquire and/or develop a comprehensive set of analytical capabilities that leverage other modeling, simulation, and HPC capabilities to ensure adequate support for future campaign endeavors.

Electronic Warfare

The electronic warfare (EW) team has demonstrated its understanding of future threats for operations in complex electromagnetic environments. With new threats, the level of analysis and assessment needed will require more nuanced applications of EW, and related activities of electronic surveillance, electronic attack, and electronic protection along with countermeasure and counter-countermeasure developments. The team has state-of-the-art laboratories and has developed state-of-the-art instrumentation, which have been integrated with modeling and simulation tools to assess and test Army radio frequency and electro-optical systems. In particular, the anechoic chamber provides special capabilities that ARL is well positioned to take advantage of. The developed digital frequency radio memory (DRFM) module has a variety of potential applications and has been integrated into a network-controllable radio signal generation system that allows command and control of a network of signal generators to create a distributed complex radio frequency test environment. The optimized modular EW network system has a flexible architecture, with growth potential to generate future and emerging threats.

The building blocks have been developed to assemble and evolve more complex EW systems. While there is a desire to insert analysis and assessment methods into earlier stages of technology development, when and where to include these inputs into the development process has yet to be determined. The EW team needs to pursue activities that improve convergence and integration of EW and cybersecurity.

Cybersecurity

The cybersecurity team at ARL applies its expertise in cybersecurity to find areas of vulnerability in systems and then provides the systems' developers or operators with guidance to enable them to improve their security. The team provides valuable services to Army programs and organizations, and its services are in demand from various Army organizations; this has led to more demand for cybersecurity services than there is the capacity to provide those services. As a consequence, there is not enough time to carry out necessary tool development and maintenance. Nevertheless, the cybersecurity team has discovered new and previously unknown vulnerabilities (called zero-day vulnerabilities) in Army systems. These and other discoveries have enabled Army organizations to remediate vulnerabilities in developing and deployed systems. This also gives ARL the opportunity to capitalize on such discoveries by conducting root-cause analyses of discovered vulnerabilities and ensure that future Army systems are free from similar problems. However, newly discovered vulnerabilities need to be broadly investigated, and, where feasible, approaches to removing similar vulnerabilities from systems need to be created and applied broadly. To date, there have been limited engagements that required assessing embedded systems, which are likely to become more important in the future. A research effort needs to start on the cybersecurity of embedded systems, which are a major likely source of future security problems.

The cybersecurity team also needs more time and resources to develop new tools and to make current and new tools more effective. Security review practices also need to be benchmarked against peer groups from the other services and the National Security Agency in order to calibrate the quality of practices and identify opportunities for improving practices. The cybersecurity team also needs to be allowed to attend Black Hat and Defcon conferences, where state-of-the-art attacks and vulnerabilities, and the techniques used to detect them, are presented and discussed.

Complex Adaptive Systems Analysis

The goal of the complex adaptive systems and analysis (CASA) program is to develop a family of simulations and associated analysis suites to provide test beds and to support experimentation. A clear perspective of how CASA contributes and crosses boundaries within ARL and DOD needs to be articulated as soon as possible with buy-in from the relevant constituencies. The current CASA capability is neither adequate nor well positioned to engage the wide spectrum of ARL needs. A CASA program that encompasses the entire acquisition life cycle, especially long-persisting logistics and sustainment challenges, could provide an opportunity for engineering analyses to directly support the Sciences for Maneuver Campaign and the full range of force operating capabilities. The CASA program also needs to be directed toward the Army standard force-on-force models.

Current manpower levels and skills mix are insufficient in both capacity and capability to adequately support a broader CASA program as it expands across ARL. ARL needs to address the lack of data scientists and operations research analysts within the CASA program; its lack of this expertise is a conspicuous shortfall that needs to be a top priority. Furthermore, future project groups need to be supported and sustained by a nucleus of operations research expertise, including data analytics, and guiding multidisciplinary groups with skill sets that are relevant to the tasks undertaken. The vital role that operations research and operations research analysts can, and needs to, play in this activity cannot be overemphasized.

CROSSCUTTING RECOMMENDATIONS

Based on the 2015-2016 reviews whose assessment is summarized in this report, ARLTAB offers six recommendations that apply across the campaigns.

Research Portfolio

Recommendation 1. ARL should articulate a vision for each research effort, its impact on options for the Army of the future, and the exit criteria to be used to decide when to terminate a project.

Integration of Research and Systems Engineering

Recommendation 2. For each campaign, ARL should address the following:

1. **Examine how projects and programs are integrated within and across campaigns and how their findings feed into one another and into common goals and share this analysis during future reviews.**
2. **Apply systems engineering principles and processes across the life cycle of projects.**
3. **Address validation and verification across the design of experiments, modeling, tests, and analyses.**
4. **Secure military-relevant subjects for tests, experiments, and field studies involving humans.**

Interaction with Industry

Recommendation 3: ARL should undertake a systematic effort to broaden and extend its awareness of research and development activities across industry.

Research Assessment

Recommendation 4. ARL should place greater emphasis and focus on a systematic assessment of its research portfolios.

Staff Development, Retention, and Mentoring

Recommendation 5. ARL should develop a structured program for professional development of its research staff and assess the program for its effectiveness.

Facilities and Equipment

Recommendation 6. ARL should complete formulation of 5-, 10-, 15-, and 20-year strategic plans linked to the campaign technical goals and objectives for facilities and capital equipment. These strategic plans should also include strategic and tactical plans for necessary computing resources, in particular, those needed to support classified computational needs.

1

Introduction

This introductory chapter describes the biennial assessment process conducted by the National Academies of Sciences, Engineering, and Medicine's Army Research Laboratory Technical Assessment Board (ARLTAB). It then describes the preparation and organization of the report, the assessment criteria, and the approach taken during the report preparation.

THE BIENNIAL ASSESSMENT PROCESS

The ARLTAB is guided by the following statement of task:

> An ad hoc committee to be named the Army Research Laboratory Technical Assessment Board (ARLTAB), to be overseen by the Laboratory Assessments Board, will be appointed to continue the function of providing biennial assessments of the scientific and technical quality of the Army Research Laboratory (ARL). These assessments will include findings and recommendations related to the quality of ARL's research, development, and analysis programs. While the primary role of the ARLTAB is to provide peer assessment, it may offer advice on related matters when requested by the ARL Director. The ARLTAB will provide an interim assessment report at the end of Year 1 of each 2-year assessment cycle and a final assessment report biennially. The ARLTAB will be assisted by up to seven separately appointed panels that will focus on particular portions of the ARL program. Each year, up to three additional panels may be appointed to assess special topics, at the request of the ARL Director.

The charge of ARLTAB is to provide biennial assessments of the scientific and technical quality of the Army Research Laboratory (ARL). These assessments include the development of findings and recommendations related to the quality of ARL's research, development, and analysis programs. ARLTAB is charged to review the work in ARL's science and technology (S&T) campaigns (Materials Research, Sciences for Lethality and Protection, Information Sciences, Computational Sciences, Sciences for Maneuver, Human Sciences, and Analysis and Assessment) but not the work of the Army Research Office

(ARO), a key element of the ARL organization that manages and supports basic research; however, all ARLTAB panels receive reports of how the research and development (R&D) activities of ARO and ARL are coordinated.

In addition, at the discretion of the ARL director, the ARLTAB reviews selected portions of the work conducted by the collaborative technology alliances (CTAs) and cooperative research alliances (CRAs). Although the ARLTAB's primary role is to provide peer assessment, it may also offer advice on related matters when requested to do so by the ARL director; such advice focuses on technical rather than programmatic considerations. To conduct its assessments, the ARLTAB is assisted by six National Academies' panels, each of which focuses on one or more of ARL's S&T campaigns. ARLTAB's assessments are commissioned by ARL itself rather than by one of its parent organizations.

For this assessment, the ARLTAB consisted of seven leading scientists and engineers whose collective experience spans the main topics within ARL's scope. Six panels, each of which focuses on one or more of ARL's S&T campaigns, report to the ARLTAB. Six of the ARLTAB members serve as chairs of these panels. The panels range in size from 10 to 31 members, whose expertise is carefully matched to the technical fields covered by the areas that they review. Selected members of each panel attend each annual review. In total, 102 members participated in the reviews that led to this report. All panel and ARLTAB members participate without compensation.

The National Academies appointed the ARLTAB and panel members with an eye to assembling a slate of experts without conflicts of interest and with balanced perspectives. The experts include current and former executives and research staff from industrial R&D laboratories, leading academic researchers, and staff from the Department of Energy national laboratories and federally funded R&D centers. Thirty-two of them are members of the National Academy of Engineering, three are members of the National Academy of Sciences, and four are members of the National Academy of Medicine. A number have been leaders in relevant professional societies, and several are past members of organizations such as the Army Science Board and the Defense Science Board. ARLTAB and its panels are supported by National Academies' staff, who interact with ARL on a continuing basis to ensure that ARLTAB and the panels receive the information they need to carry out their assessments. ARLTAB and panel members serve for finite terms, generally 4 to 6 years, so that viewpoints are regularly refreshed and the expertise of the ARLTAB and panel members continues to match ARL's activities. Biographical information on ARLTAB members appears in Appendix B.

In 2015 and 2016, the six panels reviewed the following S&T campaigns of ARL:

- Panel on Ballistics Science and Engineering: Sciences for Lethality and Protection and Analysis and Assessment;
- Panel on Human Factors Science: Human Sciences and Analysis and Assessment;
- Panel on Information Science: Information Sciences and Computational Sciences;
- Panel on Materials Science and Engineering: Materials Research;
- Panel on Mechanical Science and Engineering: Sciences for Maneuver; and
- Panel on Assessment and Analysis: Analysis and Assessment.

This biennial report summarizes the findings of the ARLTAB from the reviews conducted by the panels in 2015 and 2016 and subsumes the 2015-2016 interim report.[1]

[1] National Academies of Sciences, Engineering, and Medicine, 2016, *2015-2016 Assessment of the Army Research Laboratory: Interim Report,* Washington, D.C.: The National Academies Press.

PREPARATION AND ORGANIZATION OF THIS REPORT

The current report is the ninth biennial report of ARLTAB. Its first biennial report was issued in 2000; annual reviews were issued in 1996, 1997, 1998, 2013, and 2015. As with the earlier reviews, this report contains ARLTAB's judgments about the quality of ARL's work (Chapters 2 through 8 focus on the individual S&T campaign areas, and Chapter 9 provides a discussion of crosscutting issues across all of ARL). The rest of this chapter explains the rich set of interactions that supports those judgments.

The amount of information that is funneled to the ARLTAB, including the evaluations by the recognized experts who make up ARLTAB's panels, provides a solid foundation for a thorough peer review. This review is based on a large amount of information received from ARL and on interactions between ARL staff and the ARLTAB and its panels. Most of the information exchange occurs during the annual meetings convened by the respective panels at the appropriate ARL sites. Both at scheduled meetings and in less formal interactions, ARL evinces a very healthy level of information exchange and acceptance of external comments. The assessment panels and ARLTAB engaged in many constructive interactions with ARL staff during their annual site visits in 2015. In addition, useful collegial exchanges took place between panel members and individual ARL investigators outside scheduled meetings as ARL staff members sought clarification about panel comments or questions and drew on panel members' contacts and sources of information.

Each panel's review meeting lasted about 2.5 days, during which time the panel members received a combination of overview briefings by ARL management and technical briefings by ARL staff. Prior to the meetings, panels received extensive materials for review, including selected staff publications.

The overview briefings brought the panels up to date on the broad scope of ARL's scientific and technical work. This context-building step was needed because the panels are purposely composed of people who—while experts in the technical fields covered by ARL's S&T campaigns that they reviewed—were not engaged in collaborative work with ARL. Technical briefings for the panels focused on R&D goals, strategies, methodologies, and results of selected projects at the laboratory. Briefings were targeted at coverage of a representative sample of each of ARL's S&T campaigns over the 2-year assessment cycle. Briefings included poster sessions that allowed direct interaction among the panelists and staff of projects that were not covered in the briefings.[2]

Ample time during both the overview and the technical briefings was devoted to discussion, which enabled panel members to pose questions and ARL staff to provide additional technical and contextual information to clarify panel members' understanding. The panels also devoted sufficient time to closed-session deliberations, during which they developed findings and identified important questions or gaps in panel understanding. Those questions or gaps were discussed during follow-up sessions with ARL staff so that the panel was confident of the accuracy and completeness of its assessments. Panel members continued to refine their findings, conclusions, and recommendations during written exchanges and teleconferences among themselves after the meetings.

In addition to the insights that they gained from the panel meetings, ARLTAB members received exposure to ARL and its staff at ARLTAB meetings each winter. The 2015 and 2016 ARLTAB meetings refined elements of the assessment process focused on ARL's S&T campaigns, including read-ahead materials, review agendas, and expertise required within the panels.

---

[2] Agendas of the panel meetings can be found at the National Academies of Sciences, Engineering, and Medicine website at http://www8.nationalacademies.org/cp/ for each respective panel.

ASSESSMENT CRITERIA

During the assessment, the ARLTAB and its panels considered the following questions posed by the ARL director:

- Is the scientific quality of the research of comparable technical quality to that executed in leading federal, university, and industrial laboratories both nationally and internationally?
- Does the research program reflect a broad understanding of the underlying science and research conducted elsewhere?
- Does the research employ the appropriate laboratory equipment and/or numerical models?
- Are the qualifications of the research team compatible with the research challenge?
- Are the facilities and laboratory equipment state of the art?
- Are programs crafted to employ the appropriate mix of theory, computation, and experimentation?

To assist ARL in addressing promising technical approaches, the ARLTAB also considered the following questions:

- Are there especially promising projects that, with improved direction or resources, could produce outstanding results that can be transitioned ultimately to the field?
- Are there promising outside-the-box concepts that could be pursued but are not currently in the ARL portfolio?

Within the general framework described above, the ARLTAB also developed, and the panels selectively applied, detailed assessment criteria organized in the following four categories (Appendix C presents the complete set of assessment criteria):

1. *Project goals and plans*. Criteria in this category relate to the extent to which projects address ARL strategic technical goals and are planned to effectively achieve the stated objectives;
2. *Methodology and approach*. Criteria in this category address the appropriateness of the hypotheses that drive the research, of the tools and methods applied to the collection and analysis of data, and of the judgments about future directions of the research;
3. *Capabilities and resources*. Criteria in this category relate to whether current and projected equipment, facilities, and human resources are appropriate to achieve success of the projects; and
4. *Scientific community*. Criteria in this category relate to cognizance of and contributions to the scientific and technical community whose activities are relevant to the work performed at ARL.

APPROACH TAKEN DURING REPORT PREPARATION

This report represents ARLTAB's consensus findings and recommendations, developed through deliberations that included consideration of the notes prepared by the panel members summarizing their assessments. ARLTAB's aim with this report is to provide guidance to the ARL director that will help ARL sustain its process of continuous improvement. To that end, the ARLTAB examined its extensive and detailed notes from the many ARLTAB panel and individual interactions with ARL during the 2015-2016 period. From those notes, it distilled a shorter list of the main trends, opportunities, and challenges that merit attention at the level of the ARL director and his management team. The ARLTAB used that list as the basis for this report. Specific ARL projects are used to illustrate these points in the following chapters when it is helpful to do so, but the ARLTAB did not aim to present the director with a detailed account of interactions with bench scientists. The draft of this report was subsequently honed and reviewed according to the National Academies' procedures before being released.

The ARLTAB applied a largely qualitative rather than quantitative approach to the assessment. The approach of ARLTAB and its panels relied on the experience, technical knowledge, and expertise of its members, whose backgrounds were carefully matched to the core technical competency areas in which ARL activities are conducted. The ARLTAB and its panels reviewed selected examples of the scientific and technological research performed by ARL; it was not possible to review all ARL programs and projects exhaustively. Given the necessarily nonexhaustive nature of the review process, the omission of mention of any particular program or project should not be interpreted as a negative reflection on the omitted program or project.

ARLTAB's goal was to identify and report salient examples of accomplishments and opportunities for further improvement with respect to the technical merit of ARL work and specific elements of ARL's resource infrastructure that are intended to support the technical work. Collectively, these highlighted examples for each ARL S&T campaign are intended to portray an overall impression of the laboratory while preserving useful mention of suggestions specific to projects and programs that the ARLTAB considered to be of special note within the set of those examined.

REPORT CONTENT

This chapter discusses the biennial assessment process used by the ARLTAB and its six panels. Chapters 2 through 8 provide detailed assessments of each of the ARL S&T campaigns reviewed during the 2015-2016 period. Chapter 9 presents findings common across multiple S&T campaigns. The appendixes provide ARL's S&T campaigns and their mapping to the technical areas reviewed in 2015 and 2016, biographical information on the ARLTAB members, the assessment criteria used by ARLTAB and its panels, and a list of acronyms found in the report.

2

Materials Research

The Panel on Materials Science and Engineering at the Army Research Laboratory (ARL) conducted its review of ARL's programs in biological and bioinspired materials, energy and power materials, and engineered photonics materials at Adelphi, Maryland, on June 10-12, 2015, and its review of ARL's programs in high strain rate and ballistic materials, structural materials, and electronic materials at Aberdeen Proving Ground, Maryland, on May 24-26, 2016. This chapter provides an evaluation of that work, recognizing that it represents only a portion of ARL's Materials Research Campaign.

ARL's materials research spans the spectrum of technology maturity and addresses Army applications, working from the state of the art to the art of the possible—25 years into the future—according to ARL. Materials research efforts and expertise are spread throughout the ARL enterprise. As the ensemble of the materials discipline and capabilities, materials sciences is one of ARL's primary core technical competencies. The materials sciences work supports the mission of ARL, as the U.S. Army's corporate laboratory, to provide innovative science, technology, and analyses to enable a full spectrum of operations.

The Army's mission is fundamentally intertwined with its ability to produce new and improved materials with a combination of properties that are more often than not unique to the requirements of the warfighter. As such, materials science is one of ARL's core technical competencies, with the Materials Research Campaign spread throughout the ARL enterprise.

BIOLOGICAL AND BIOINSPIRED MATERIALS

The scientific quality of the work in this area is on par with that of leading federal, university, and industry laboratories, reflects a broad understanding of the underlying science and research being conducted elsewhere, and is recognized as a component of the broader national effort in biomaterials research through its government (e.g., the U.S. Army Edgewood Chemical Biological Center and the

U.S. Army Natick Soldier Systems Center) and university (e.g., Institute for Collaborative Biotechnology at the University of California, Santa Barbara) partnerships and collaborations.

This research area has grown substantially over the last 2 years. The knowledgeable leadership direct principally competent, early-career scientists. Relative to the importance of this research, the personpower in this area is considered suboptimal.

The laboratories are generally well equipped to perform the types of studies and analyses required for the biological research. The biological characterization and imaging tools are good, with recent additions of next-generation sequencing and protein synthesis and medium-scale bioreactors with real-time metabolism analysis capabilities. A peptide sequencer would provide important missing capabilities and would help to accelerate research.

Among the excellent research activities of this group, particular promise is shown by the stabilization of proteins against thermal and chemical extremes, using new chemistries and methods to derive antibody-like reagents that improve on antibody properties (specifically, biomolecular recognition and binding characteristics).

Accomplishments and Advances

Biomaterials for Hazardous Materials Detection

The development of synthetic bioreceptors as alternatives to antibodies is being pursued to allow biosensing outside the laboratory and in conditions more representative of those found in the field (e.g., high temperature). Overall, this project, conducted by a relatively new group of researchers, is equivalent to the best work performed elsewhere. The work supports a variety of missions, including water and food defense, individual soldier protection, and collective protection. The concept is to mimic antibody binding with small peptides. An approach screens large libraries of enhancing affinity, selectivity, and other desired features (e.g., serum stability) via an iterative process. The group is exploring a number of different strategies to perform this screening, which is a significant strength of its approach.

To create high-affinity and robust biosensors, independently binding peptides are chemically conjugated using click chemistry to identify bi- or greater ligands. The use of cyclic peptides in place of linear peptides is also being explored as a means to higher-affinity molecules. It is impressive that this technology has allowed rapid (less than 1 week) identification of binding peptides. The group's demonstration of binding to aluminum alloys provides a practical example of its capabilities. Overall, this very productive group is doing cutting-edge work that complements work ongoing in extramural laboratories.

Given the alternative strategies for achieving similar outcomes (e.g., single-chain thermostable antibodies), more specific performance criteria or target product profiles will be needed for further development of some of these areas.

Biohybrid Materials for Sensing

Bio-nano-hybrid systems are being investigated for their potential applications for in vivo physiological monitoring, nanomedicine, traumatic brain injury (TBI) dosimetry, and other photonics-based sensing. The intent of this research is to understand the interactions taking place at the biomediated, nanocrystalline, photonic or nanophotonic biomaterials interface, and to develop new designs for tailored light or matter interactions that can be applied to Army needs. The examples presented use proteins to stabilize nanoclusters and control photonic materials properties. Protein-nanocrystalline structures have been embedded with neurons to detect primary blast-induced neurotrauma, a potential means to investigate mild TBI. The protein-stabilized nanoclusters (P-NCs) were synthesized in situ in neuronal and nontumorigenic cells—the first demonstration of in situ nanocluster growth in nontumorigenic cell lines.

This project provides a good example of grassroots-driven collaboration with outside laboratories. It is very good fundamental research with potential applications to sensors (ligand recognition) and to cell targeting for drug discovery and development. The research is characterized by good integration of modeling and experimental work across bio-molecular- to cellular-length scales. However, more fundamental work is needed to determine the location of nanoclusters, to determine whether protein(s) stabilize the nanoclusters, and to validate in vitro expressions under high pressure.

The effort to utilize P-NCs for monitoring pressure in TBI appears to yield distinct spectral peak intensity changes. Without concurrent modeling efforts it is not certain whether these intensity differences can be due solely to changes in nanocrystal clusters in proteins or can also be due to other effects. It is therefore unclear whether this research would be better directed toward sensors for TBI or other extreme conditions.

A smaller project is focused on the development of a real-time handheld detector for synthetic cannabinoids based on use of a cannabinoid receptor as a transduction element for detection of contraband material. This is an attractive approach, given the diversity of targets (generated by the illegal synthetic drug community) that can be detected using the functional receptors that trigger downstream cognitive effects. Further investment in this project is expected to depend on performance parameters that are yet to be determined, including the limits of detection, the dose response across a useful operational range, and the signal to noise performance in the presence of interference.

Bioinspired and Biomimetic Materials for Protection

This effort addresses a number of related topics intended to improve the performance of polymers in areas relevant to the Army mission, as well as the use of polymers in studies of TBI. ARL's biobased polymer program has been used to produce transitioned biorubber toughening agents, reactive diluents, monomers for polyamides, biobased bisphenol A analogues, and multiphenolic monomers. One program was directed toward developing high-performance biobased polymers for Army applications. The goal of the program is to utilize renewable lignin-based resources to create molecules for the production of high-performance polymers. Successes to date include synthesis of monomers of diepoxy and demonstration of polymers with very high glass transition temperatures. The associated challenges include development of scalable chemistries and structure-property-toxicity capabilities that would allow for transition of the technology to industrial partners.

Another project focused on improving the properties of polymers by incorporating reversible cross-links to enhance toughness. The goal of a third project is to develop high-temperature adhesives that are inspired by the extraordinary properties of spider silk and muscle titin, specifically by incorporating reversible cross-links whose breakage can allow the unfolding of polymer domains. While inspired by natural polymers that derive their mechanical properties from hydrogen bonding, this project focuses on the use of reversible metal bonds as high-temperature tougheners. The emphasis in both projects on developing a mechanistic understanding is a strength, because so much of the other work on these topics is empirical. The emphasis on high-temperature performance differs from the focus in most other laboratories that work on these topics. It is unclear, however, whether simple insertion of metallic elements will achieve the strength and toughness levels of the natural polymers, but new insights are likely, and the coordination with modeling is laudable.

A small project presented in this area addresses the extremely important problem of TBI. The experimental setup devised is fairly simple: A small (2 g) explosive charge is detonated close to a tank that contains neuron cells. This approach to load cells is novel compared to other TBI studies using cell cultures and may yield new insights. Although this is an exciting project, it needs to be part of a much larger and broader program studying TBI; it could be connected to efforts taking place at other Department of Defense (DOD) laboratories. This project needs to also consider more interpretable dynamic loading of neuron cells. Test configuration could allow simulation of pressure waves so that any observed changes in the neuron cells can be related to a known pressure history.

Bioconversion, Biosourced Energy

This is an appropriately focused long-term effort to address Army-specific needs for dealing with food and water wastes. It connects well with other Army entities and is appropriately resourced in terms of both equipment and competent personnel.

A positive characteristic of the program is its university outreach and collaborations, intended to draw in expertise and technologies. These relationships may be leveraged or enhanced through the developing ARL open campus initiative. A more formal connection with the Army Medical Command, particularly in the areas of wastes and health, will be important.

To achieve the programs' goals, it may be worthwhile to put more emphasis on high-throughput approaches to empirically screening large numbers of communities; this could allow more rapid identification of desired bacterial communities. It might also prove useful to examine lessons that may be learned from the limitations of previously fielded systems.

Opportunities and Challenges

Because biology is a growth area, ARL has an opportunity to identify and recruit a critical mass of biologists, including microbiologists and polymer/organic chemists, looking well into the future to create an integrated community of researchers. The process of recruiting and retaining talent could encourage better articulation of the expectations and career paths that lead from postdoctoral researchers who are contractors to scientists who are government employees, and to develop an effective mentorship program emphasizing professional development and job satisfaction.

ARL needs to reexamine its polymer-related work to assure the closest possible relations between the researchers at its Adelphi and Aberdeen locations.

ENERGY AND POWER MATERIALS

Accomplishments and Advances

The quality of the research projects, the staff, and the facilities is comparable to high-quality research laboratories elsewhere in industrial and academic environments. Where there are gaps in the technical skills or methods needed for a project, the ARL staff demonstrate mature experience and judgment in seeking out high-quality collaboration with other non-ARL researchers within and beyond the Army research enterprise.

The early-career researchers are strong and have excellent skills, which likely reflect good mentorship by senior personnel. Importantly, the research staff are enthusiastic and throughout the review demonstrated a clear focus on Army needs, an appreciation for the importance of moving basic research to technology to impact, and skill in selecting research methods and tools involving experiment, theory, and simulation.

The portfolio of research projects reviewed included an appropriate balance of high-risk, long-term-impact projects along with mid-term and short-term projects. There was a broad, deep coverage of different devices, different fuels, and different applications covering a wide range of size and time scales.

There are continuing improvements in research quality, staff hiring in both postdoctoral and permanent positions, and collaborative activity. As part of these improvements, ARL has expanded its modeling capabilities. The current in-house capability for carrying out high-level simulation and modeling activities is of high quality and moving in the right direction.

Advanced Energy Storage: Advanced Battery Chemistry

Though the advanced battery effort at ARL is small relative to similarly focused programs at other federal laboratories (e.g., Department of Energy laboratories), it is internationally recognized for its high scientific quality and long history of productivity and innovation.

The research includes significant elements of experimental and computational numerical modeling work. The laboratory equipment for experimental work is excellent, spanning an impressive range of capabilities from materials synthesis and characterization, to electrode and cell fabrication. Computational efforts in the battery area are good but appear to be relatively recent. They could possibly benefit from additional resources and emphasis.

The team has excellent qualifications that are well matched to their research challenges. In addition, program participants have an excellent understanding of research conducted elsewhere and are well aware of critical research issues and advances from around the world. The ARL team has formed a local Center for Batteries in Extreme Environments, which provides a good model for interaction of ARL staff with non-ARL scientists.

The projects in this area also are synergistic with one another. The team thinks hard about transition pathways to scale-up, manufacture, and commercialization and seems well positioned to make decisions and negotiate arrangements to transition ideas to the field.

The team has begun generating a database of properties on electrolytes, including interfacial reactivity. A pathway may exist for organizing these data in a manner similar to that being pursued for other battery materials in the materials genome initiative. Productive work may come from the team's interactions with that initiative focused on electrolytes, possibly including the effect of additives on interfacial reactivity.

Advanced Energy Storage: Structural Batteries Using Additive Manufacturing

The researchers are successfully developing techniques for fabricating multifunctional battery materials using additive manufacturing (AM). The lattice structures constructed using AM have favorable mechanical properties and controllable surface area per unit volume, which permits tailoring and optimization of electrochemical performance. With respect to weight reduction for batteries and capacitors, the AM method has clear advantages over earlier methods. The measured elastic properties and electrical performances of the fabricated materials agree well with the finite-element modeling performed as part of the project.

The techniques and materials are promising, and the scientific quality of the project is comparable to quality at leading research institutions. A more comprehensive modeling and simulation component addressing chemistry and physics, in addition to mechanics, might be desirable for understanding effects such as the influence of porosity on performance of gels. The project has a good balance of theory and experimentation. It would be good for the team to consider the previous work on nanotrusses done at the Naval Research Laboratory to see if there is anything in this work that might be applicable. To help design for sufficient mechanical durability and reliability of materials, it might be worthwhile to investigate the strength and failure properties of the fabricated materials as well as their elastic moduli. It may also be desirable to consider the practical issues associated with scaling up the laboratory AM process to full-size batteries. This project has significant potential for innovative discovery.

Alkaline Fuel Cells: Optimizing Structure and Chemistry of Ion-Containing Polymers for Charge Transport

This project focuses on the development of alkaline fuel cells. Conventional approaches rely on a liquid KOH electrolyte as a means of OH^- exchange. This electrolyte is problematic because it is a liquid and can be poisoned with CO_2 owing to the formation of K_2CO_3. The goal of this project is to circumvent these problems using a polymer electrolyte. In particular, a mixture of dicyclopentadiene and CO is used to create a bicontinuous microstructure intended to maintain high OH^- conductivity and strong mechanical properties. The OH^- conductivities achieved were the best reported, though the mechanical behavior was not adequate. By increasing polymer molecular weight and cross-linking, strength and toughness were increased nearly twofold, but this is still far less than competing materials. It is unclear whether this mechanical behavior would be acceptable. The researchers were able to enhance material behavior by optimizing microstructure, which was in turn achieved by increasing the connectivity of the hydrophilic domains.

This research is Army-relevant, reflects a correct understanding of the literature, makes use of state-of-the-art facilities, and is of high quality, comparable to similar university and industrial efforts. However, the scope of the effort requires expansion if the work is to have a substantial impact within this community. Additionally, being more engaged with this community—e.g., the multiuniversity research initiative at the Colorado School of Mines—would help this work proceed by enhancing critical decision making—for example, in materials selection. This project would be strengthened significantly by the addition of a modeling component. There are many existing methods and codes that could be helpful; for example, the group at the National Renewable Energy Laboratory is modeling this sort of system.

The publication record of the researchers is prolific: Seven papers have been published, one is under review, and four more are in preparation.

Alternative Energy Photovoltaics

This project utilizes quantum dot nanomaterials for photovoltaic conversion and focuses on enhanced light absorption and minimizing reflective losses. This nanomaterial approach eliminates the need for a traditional tracking system and, if successful, would significantly impact a number of Army applications requiring flexible and efficient power.

The researchers have established productive collaborations with the communities at the University of Texas, the State University of New York, Microlink Devices, and the University of Michigan—all characterized by a good mix of experiment, theory, and simulation. The researchers demonstrate a broad understanding of the related science as exemplified by their efforts to modify the wetting layer thickness to increase electronic capture; they have achieved photovoltaic (PV) efficiency 6 percent above the record for GaN.

Alternative Energy: Highly Mismatched Alloys

This project develops material to split water by using sunlight as the energy source. This research is high risk but potentially offers a very high payoff. The idea, based on results appearing in the literature, is to replace some N with Sb in GaN to form GaN_xSb_{1-x}. It was predicted that by adding Sb, the bandgap could be lowered to about 2.2 eV, producing an efficient light absorber. These alloys, highly mismatched in size or electronegativity, have never before been synthesized. The group has significant experience with Group V alloys and apparently a unique synthesis capability.

The experimental results shown verify the bandgap crossing model (developed at Lawrence Berkeley National Laboratory [LBNL]) up to $x = 0.22$. Materials produced remain crystalline. It appears that very small amounts of Sb lower the bandgap significantly. The principal investigator did not understand why the model predicted such behavior. An understanding of the controlling physics needs to be developed.

Attempts could be made to fabricate a device, though this necessitates doping these materials. Doping can now be done for GaN, but it is not clear what effect the Sb will have on this process.

Good collaborations with LBNL and with the University of Strathclyde and the University of Nottingham (both in the United Kingdom) are ongoing. Only theory is done at LBNL. There is also significant competition from the National Renewable Energy Laboratory and the University of North Carolina; these groups are focused on different materials. This project could be aided by more modeling. There have been five publications by this group in the last year.

Alkaline Fuel Cells

The principal objective of this research is to create anion exchange membrane/proton exchange membrane stacks, eliminating the need to transport water throughout the cell and potentially reducing the mass and footprint of the device. The principal investigators have pulled together an excellent team, including a group at Georgia Institute of Technology.

The principal investigators were aware of many of the relevant issues for successfully constructing such a cell, including issues regarding delamination. This work has a robust modeling component. Overall, this research is innovative and promising.

Alloy Type Anodes for Lithium-Ion Batteries

Lithium (Li)-ion battery performance and weight reduction may be improved by using silicon (Si) anodes to increase the capacity for Li storage. This project addresses an important practical difficulty with Si anodes, which is their tendency to experience mechanical failure after a small number of electrical discharge and recharge cycles. The principal investigator has performed careful in situ measurements of this effect using an atomic force microscopy technique. The implementation of these experimental techniques is the main achievement of the work so far. The principal investigator is also working on coatings for anodes to reduce cracking, with encouraging results. The project also supports collaborations with the University of Utah to use molecular dynamics (MD) simulations to study the electrical and mechanical processes involved. MD seems to be a very promising method for understanding the fundamental aspects of the cracking problem. In particular, analysis might help to reveal why thin coatings apparently reduce damage in spite of the very large linear strains to which the coatings are subjected. The project would benefit from a closer working relationship between the experimental and computational team members.

Beta(photo)voltaics

The project is intended to develop a long-lived (25-year goal) power source using beta and alpha energy conversion in wide bandgap (WBG) semiconductor materials and phosphors. The approach uses a beta emitter (tritium) to produce electricity. Current designs do not produce enough power to be useful for the Army, and so the isotope power source is coupled with a Li-ion battery to take care of power demands during higher current demands such as during signal generation. The isotope power source is used as a trickle charger for the Li-ion battery. Electrochemical capacitors were used, but there was high leakage current.

This project started as an engineering problem. Isotope power sources have been used for years in weapons applications. Since the Army has access to isotope materials, it made sense to utilize this approach for power production. To test the concept, an isotope power source was fabricated that generated 100 μW.

The innovative concept applies to the investigation of three-dimensional (3D) interaction space in WBG materials and phosphors to increase energy conversion and efficiency. The project is well thought out, and it has a high probability of success. The principal investigator is performing mechanical cross-section simulations to aid in design, so the mix of theory, experimentation, and computation is sufficient. The qualifications of the researcher and the facilities appear to be compatible with this research challenge. If successful, this could open a myriad of small power source applications for the Army.

Carbon Formation During Catalytic Oxidation of Hydrocarbon and JP-8 Fuel

The use of logistics fuel for compact, heat-driven electric power generation is compromised because sulfur impurities poison the catalytic activity of microcombustors. In this project, a materials-by-design approach is being used to identify promising combustion catalysts, which are investigated with experimental and computational methods. In situ spectroscopy is incorporated with short contact-time reactors to identify surface species during catalytic combustion of prototype fuel, while simultaneously monitoring poisoning. These data, used in conjunction with a microscopic reaction diffusion model of surface events during combustion, clarify the effect of sulfur. It has been recognized that sulfur enhances carbon formation on platinum (Pt) but not on rhodium (Rh). The project promises to accelerate microcombustor catalytic design through reactive flow modeling. This is good scientific work linked with sound engineering methods for scale-up and extension to logistic fuels.

Critical Solvation Issues in Lithium-Ion Batteries

This poster describes part of the excellent battery program that focuses on Li salt solvation in organic carbonate and water solvents. The objective is to better understand fundamental electrolyte interface properties. The work showed preferential solvation of Li by ethylene carbonate (EC) in EC/dimethyl carbonate mixtures, which is relevant to solid-electrolyte interphase formation at carbon anodes. New water-in-salt electrolytes having less than 20 weight percent water in Li bis-trifluoromethanesulfonimide were prepared, and early-stage results are quite promising. This is an example of an emerging class of electrolytes called deep eutectics. Water in this electrolyte is thought to have very different properties from conventional water because such a large fraction of these water molecules are contained in solvation shells. Understanding of interface passivation could help in electrolyte material choice. The work is led by an energetic PI and is synergistic with the overall thrust of the ARL battery effort. Its high scientific quality is demonstrated through publication in quality journals.

High-Voltage Li-Ion Electrodes and Electrolytes

This project involves work on olivine $LiMPO_4$-type cathodes, where M is Co, Mn, and Ni substituting for the usual Fe. These three metals have more positive redox potentials, so batteries using these materials have higher voltages. $LiCoPO_4$ has a high potential but usually exhibits significant capacity fade upon cycling. The ARL team found that mixing some Fe with the Co results in much less capacity fade, with minimal loss of overall capacity. A mechanistic understanding of the diminished capacity fade is being pursued. The work is of high quality and fits well with the significant worldwide effort to identify new battery cathodes to enable higher energy density batteries.

Isomeric Materials Research

This project addresses the use of nuclear transitions for energy-on-demand. The concept is to convert a long-lived, excited nuclear state to a short-lived ground state by excitation by photons or neutrons. The main scientific content is nuclear physics, in contrast to the mainstream atom/electron/photon-centered work in the ARL Materials Research Campaign. The work requires investigation of level diagrams for candidate nuclei. Because of the complexity of the few-body problem for large nuclei, the nuclei cannot be modeled to sufficient accuracy but have to be measured. The investigators combine information from the literature with their measurements. The conversion concept has been demonstrated for a silver isotope. The work is of high technical quality and is published in the appropriate journals. This is a long-term, high-risk approach with regard to practical applications, with many questions to be answered, including how to produce the long-lived excited state in sufficient quantities, but it is worth pursuing. If successful, impact could be high.

Lattice Conductivity of Dense Ta-Doped $Li_7La_3Zr_2O_{12}$

LLZO is a candidate for a solid electrolyte in Li batteries. The goal of this study is to enhance Li conductivity via doping Ta for Zr. Using this approach, the investigators were able to reproduce a similar study by Goodenough and suggested that other enhanced results in the literature were likely due to other defects. This is a very good addition to the literature, where reproducibility of key results is often lacking. Furthermore, this work had a strong density functional theory component that was performed at the Naval Research Laboratory (NRL). The work would be outstanding if there were researchers within ARL who could complement the efforts of NRL.

Mathematical Modeling and Lifetime Extension of Thermal Batteries

Thin-film thermal batteries could provide improved reliability and performance over present designs in munitions. This project is successfully developing a comprehensive analysis tool that models the thermal energy balance, gas generation, and electrical performance of thin-film thermal batteries. So far, it is mainly the thermal problem that has been addressed. The ARL thermal model has been integrated into Sandia National Laboratories' Sierra finite element code. The method that ARL is developing could be combined with a mathematical optimization tool, providing a direct and systematic way to improve battery design. There was no mention of validation of the model, suggesting that this is not a central focus. The principal investigator did not seem to understand the model being used. Therefore, in addition to adding the multi-physics capabilities that are planned, it would be desirable to obtain experimental data that would be needed to validate the submodels (e.g., thermal, gas transport, electrical, chemical) as the code grows in size and complexity.

Pyroelectric Materials for Energy Applications

Led by a competent postdoctoral researcher, this project draws on a 2006 paper[1] reporting a giant electrocaloric effect in perovskite oxide PZT (metallic oxide based piezoelectric material) to propose and prototype a device to convert heat/infrared photons to electrical energy. The concept is to run a heat engine loop between the low polarization branch of the polarization versus field (P-E) hysteresis loop at high temperature and the high polarization branch at low temperature. There is much room for optimization by choice of materials and by controlling the quality of the thin film, with leakage current being of particular concern. The collaboration with the University of California at Berkeley will help with the latter. The long-term application will be the remote supply of energy by an infrared laser and could be used, for example, to recharge drones in flight. The work is of high technical quality, and the project leader communicates well with the broader community. As the work progresses, more modeling could be integrated into the project.

Understanding C-C Bond Breakage on Plasmonic Nanostructures

This proof-of-concept project is directed toward developing catalyst structures for breaking the C-C bonds associated with high energy density logistic fuels (e.g., ethanol) using light-harvesting nanoscale arrays formed by localized surface plasmon resonance. This is a unique fabrication approach based on a good concept for photo-reformation of logistic fuel. The project represents a high-risk endeavor. The first steps toward forming such structures have been made with use of the Specialty Electronic Materials and Sensors Cleanroom facility. Initial work includes modeling the plasmonic aspects of the structure. Additional modeling worked is planned to take account of the immersed, reactive, electrochemical environment. Experimental characterization of the photo-induced reactions is planned, using well-established electrochemical and surface science methods. Desorption mass spectroscopy electrophotometry has been reported by others and may be considered for this project. At this point, there are no experimental data on the structure, and the current understanding of the reaction mechanism is speculative. It would be helpful at this point to create a device and test it out. It is not yet clear whether this fabrication method can be implemented at large scale.

Grain-Boundary Engineering of Ion-Conducting Ceramics

This project examines two approaches to reducing ion transfer resistance at grain boundaries in the fast lithium-ion-conducting ceramic $Li_{3x}La_{2/3-x}TiO_3$, $x = 0.11$. Previous work showed that this solid-state electrolyte material had good conductivity. The idea is to change grain boundary (GB) properties to improve GB conductivity. Grain boundary modification was achieved by simple lithium-ion exchange from solution, and by silica coating the starting particles with SiO_2 using magnetron sputtering. Both approaches provided modest increases in conductance, both at GBs and in the bulk. A study on variations in bulk ionic conductivity with thermally induced changes in crystal structure was also pursued. The mechanism by which surface coatings change GB conductance is not fully understood; more structural characterization work—for example, by transmission electron microscopy (TEM)—is planned to help address this point. It was also found that different processing conditions could change crystal structure and conductivity. The principal investigator, a postdoctoral researcher with ceramics experience relevant

[1] A.S. Mischenko, Q. Zhang, J.F. Scott, R.W. Whatmore, and N.D. Mathur, Giant electrocaloric effect in thin-film PbZr0.95Ti0.05O3, *Science* 311:1270, 2006.

to Li battery technology, needs to learn more about the battery aspects of the work to become fully integrated into the overall effort. Still, this is excellent work for a relatively new employee. The work is of high quality and contributes to a growing body of knowledge regarding use of ceramic materials to replace liquid or polymer electrolytes in Li batteries.

Opportunities and Challenges

Questions remain as to whether ARL was mobilizing aggressively enough to capitalize on both internal advances and external advances made by the broader community—for example, whether the recent world-leading results on enhancement in quantum well infrared photodetector (QWIP) efficiencies are being translated into capability demonstrators for manufacturers and customers. Similarly, ARL may not be working to leverage external advances in silicon photonics, especially with regard to heterogeneous materials. However, in both cases, these concerns were partially allayed by discussions with staff and management regarding the status of some programs related to these technologies. For the QWIP work, for example, ARL has hired an external business consultant and is working with NASA and others on technology transfer for the QWIP breakthroughs. Nonetheless, there remains more opportunity for ARL to capitalize on its internal and external advances.

The enormous potential impact of the photonics work could have been presented more vigorously and compellingly. One way of doing so could be to augment an individual photonics presentation with an explicit description of the broader potential impact if it succeeds. Army goals were noted, but they often comprised immediate technical targets as opposed to what the ultimate impact could be for a more comprehensive field of science or for broader Army applications.

The presentation on structural batteries using additive manufacturing has significant potential associated with its innovative approach. The project combines novel fabrication methods with insight into selection of compatible multifunctional elements that combine structural components with energy storage components. Experimental work is carried out concurrent with modeling studies that guide system design choices. The external collaborations are facilitated by a flexible methodology that provides easy incorporation of next-generation subcomponent materials as they are developed. However, the effort needs to grow across a wider range of projects, with a focus on identifying appropriate modeling methods and on closing the experiment–theory–simulation loop. Increased interaction with the significant computational resources of ARL could help bridge the gap until additional capacity is available within the Materials Research Campaign. At present, first-principles computational modeling is growing, mainly through collaboration with recognized experts elsewhere, guided by very capable but limited-in-number experienced internal research staff.

In comparison to the expansion in first-principles modeling, engineering models are underutilized, perhaps because in-house expertise in this facet of modeling is limited. Engineering models are typically developed at the outset from a simple set of input parameters or components that, together with the model, predict system behavior. These components are improved as empirical knowledge of the system's behavior increases. Routine methods are now available to identify the most sensitive components for which improved fundamental knowledge is needed, to provide uncertainty quantification, and to guide system-level optimization during scale-up or scale-down beyond experimental regimes. The combination of an appropriate engineering modeling effort with the intuitive understanding of experimentalists is a highly effective engineering approach and needs to be targeted as a growth area.

In some energy and power applications, such as Li-ion batteries and fuel cells, there is a broad, vigorous, fast-moving, worldwide research effort directed toward identifying fundamental scientific issues and developing novel materials and entire systems. Accordingly, the narrowly focused ARL projects need to pick the right niche in order to have impact. The knowledge necessary to define the goals of such projects depends critically on tracking research advances elsewhere. Because postdoctoral and other early-career permanent staff researchers benefit from exposure to research activities beyond ARL, it is critically important to promote and expand active mentoring by senior staff.

ENGINEERED PHOTONICS MATERIALS

The quality of the work in photonics materials is comparable to that found at most research universities. This is an impressive accomplishment in light of the inherently wide scope of the technical program, which is essential to addressing diverse current and future Army needs. The quality of the work presented reflects a high level of technical competence and professionalism on the part of the researchers and management.

The portfolio of the engineered photonics materials group shows a good balance of high-risk, longer-term work with nearer-term customer-driven solutions or incremental, critical technology refinement. This well-balanced portfolio is supported by a strong materials capability in staff expertise and laboratory or clean room infrastructure. Investments are impressive for computational modeling and simulation that ARL has successfully implemented to complement its strengths and core competencies in materials synthesis and characterization, as well as device work. All of these facilities and capabilities are being leveraged into compelling device and application-driven work, especially in ultraviolet (UV) materials, infrared (IR) devices, and the device physics in both areas. In addition to technical diversity, there is workforce diversity.

Accomplishments and Advances

Alternative Energy: Photovoltaics

This project involves work to improve performance of low-concentration photovoltaic cells targeting robust, lightweight power for soldiers in theater. The technical focus is developing solutions using III-V quantum-dot materials to extend performance into the longer wave regions of the solar spectrum, and to improve efficiency by minimizing recombination.

Solar PV is one of the important pathways to reducing the weight of power solutions in theater. The experimental work showed solid progress, reflecting the strong competence of the team, which evinced expertise that includes epitaxy and sophisticated quantum dot engineering, polyethylene terephthalate (PET) moth eye surfaces, and intentionally induced morphological features on III-V layers for enhanced photon capture. There appeared to be extensive collaborations with researchers outside ARL.

The wetting layer state-engineering designs might benefit from more direct experimental verification of their efficacy in reducing recombination in the dots. There was a lack of clarity on the trade-offs between the high concentration (30 to 100 times what is typically seen when realizing the benefits of advanced materials) and the low concentration (less than 4 times what is typically required in nontracking applications). More clarity is needed on the system-level incremental cost of multijunction cells with significantly higher efficiencies relative to single-junction material solutions such as GaAs, which is still very high ($40,000 per square meter), or the quantum dot approach pursued in this work. Additional questions include comparisons with spectral splitting, which was examined in the DARPA-sponsored very-high-efficiency solar cell program.

Biophotonics

Progress was reported on protein-wrapped fluorescent metal nanoparticles, motivated by their potential use as neuronal pressure sensors. The long-term goal is to develop a fundamental mechanistic understanding of mild traumatic brain injury onset and development.

The fundamental work on the biomediated synthesis of atomic nanoclusters was compelling, and the fact that the proteins retain their native functionality after synthesis has tremendous potential. For example, the resulting nanoparticles may be noncytotoxic, and it may be possible to direct them to specific locations within a cell. These nanoparticle building blocks are anticipated to provide unique opportunities based on their interesting optical and physical properties. An example given was fluorescence change with pressure seen for one protein but not a different protein, an indication that interesting protein science may be enabled by this system.

There is some concern regarding the specific proposed application for these particles for understanding shock waves in tissue. The fluorescence changes with pressure were small (20 percent over 400 MPa for one system and about 6 percent over 600 kPa for a different system). In real tissue, these small changes over less than 1 ms, from a single or a few particles, will be very hard to observe. What is needed is a deeper physical analysis of the full system, including the signal-to-noise ratio in realistic shock wave and illumination conditions, and what is anticipated at a single neuron level. Also needed is a comparison with other potential techniques, such as Forster resonance energy transfer and plasmonic particles, in the context of nanoscale pressure sensors.

This ambitious work offers strong opportunities for discovery; it is a high-risk early-stage effort in ARL's expanding biophotonics effort.

Modeling and Analysis of Ultraviolet-Light-Emitting-Diode Materials

The objective of this project is to use many-body theory to model lifetime in III-nitride structures, including free carrier and exciton effects, polarization fields, and density-dependent screening of Coulomb interaction and polarization fields. This is one of the projects indicative of ARL's investments in more comprehensive modeling to support its strong core materials capabilities and competencies.

This a very challenging problem, and the principal investigator is making good progress in describing radiative lifetime, including many-body effects such as phase-phase filling, screening, and quasi-particle renormalization. However, nonradiative processes were not described at the same level of theory. Semiempirical, nonradiative models using activation energy were shown not to fit experimental data well, but improved fits were achieved with a combination of a fixed temperature-independent component plus an activation-energy component.

The development of first-principles-based and self-consistent predictive capabilities to describe carrier lifetime in III-nitride structures, including both radiative and nonradiative processes, is not easy. However, the principal investigator presented a scientific strategy to make progress toward addressing this challenge. The strategy calls for alloy fluctuations, a many-band description of the electronic wave function, the use of nonparabolic bands, and the inclusion of nonradiative recombination processes. This is a project of high technical merit and of potentially high impact in support of the Army's mission.

Ultraviolet Avalanche Photodetector Research

This work entailed the compelling development of models and experimental devices and materials to evaluate the efficacy of novel solutions for improved single-photonic avalanche detectors in the UV as replacements for photomultiplier tubes.

The principal concept is to use GaN and AlGaN epitaxial layers to address the reduction in quantum efficiencies that stems from the use of semitransparent metal electrodes on current SiC devices. Self-assembled monolayer structures were introduced to either isolate the SiC to a multiplication layer or to just use the AlGaN as a transparent contact layer to keep the SiC away from surface so as to avoid surface recombination.

This work is promising and has high-quality external partnerships. It has mainly involved epitaxy development and Si diffusion studies, and the transitioning of these to device results in avalanche operation is awaited.

Short-Wavelength Infrared Device Modeling and Optimization

This project is directed at the development of a comprehensive model that combines the finite-difference, time-domain electromagnetics of nanostructured surfaces with finite-element modeling, drift-diffusion transport to understand and optimize device designs and material structures. The model is comprehensive in that it included material, electronic, optical, and especially nanostructured geometric properties that strongly impact the electromagnetics. The integrated software suite allowed analysis of very complicated multipixel arrays, and the principal investigator showed how more simplistic models would not properly capture major performance factors. One example was that the performance of nanostructured cones could be estimated reasonably well with effective medium models at longer wavelength, but at shorter wavelengths complex scattering among the cones dominated the performance. The model was shown to be useful in assessing pixel cross talk in arrays, as well as heterostructure design and junction location for optimization of collection efficiency while minimizing generation-recombination (GR) dark current.

This is an excellent project directed toward an important topic in terms of the needs of both the Army and the broader technical community.

Diode-Pumped Tm/Ho Composite Fiber 2.1 μm Single-Mode Laser

The goal of this research is to provide a simpler and more compact 2.1 μm thulium (Tm)/holmium (Ho) source capable of achieving 100 W power in eye-safe lasers for situation awareness, monitoring, and tracking illumination and, perhaps, frequency conversion to directional IR countermeasures.

Early work has been conducted on an innovative concept to make a dual-core fiber laser that would support thulium lasing at 1,950 nm in a multimode core that would, in turn, pump a Ho single-mode core at 2.1 μm. This design is intended to achieve two excitations in the Tm with a single optical pump in the 800 nm range. This is an interesting concept, but it is too early to expect definitive evaluation of the potential.

This effort may now be positioned to benefit from a stronger modeling component to resolve the impact of saturation on spatial mode competition and laser performance. Suitable baseline modeling capabilities are readily available in the literature, and in conjunction with a more deliberate experimental plan, the modeling may be useful for isolating critical performance trade-offs.

The principal investigator is engaged in a valuable external partnership with strong competence in these fiber materials.

Thermal Property Engineering: Exploiting the Properties of Ceramic

This project consists of preliminary work on improving mid-IR lasers by increasing the effective thermal conductivity of the gain media, using nanoscale composite MgO (high thermal conductivity) with Er:Y_2O_3 (the gain media). This work addresses many scientific and engineering challenges, including the achievable effective thermal conductivity of the composite, which may be limited by phonon scattering, and the achievable volume fraction of gain media needed to be competitive with current solutions.

This work has high potential, and it may benefit from some early modeling to determine the property bounds and trade-offs. The team could also be more vigilant in reaching out to others, including the Air Force Research Laboratory, to evaluate similar work.

Photoacoustic Spectroscopy for Hazard Detection

This project involves work on an elegant and simple device approach for detecting trace elements. While many optical detection techniques are available, these are usually large and contain many precision optical elements. The detection technique proposed is small, robust, and potentially inexpensive, if applications supporting high-volume laser production are realized.

Engaging more broadly with the outside community would be beneficial, including with vendors of existing optical sensors and comparative testing on species of current interest, perhaps in the context of ARL's open campus initiative. In addition to offering a potential for more pervasive use, this will better ensure that this transitions into a product useful to the Army.

Understanding Inkjet Printed Standards for Optical Measurements

This work involves a system based on the well-tested use of inkjet printing. Although ARL has used only a single print head, the researchers have been able to print on many materials (e.g., rubber, metal, and wood) with contaminants included. The system can be used to understand how the samples age, and the flexibility of patterning and reproducibility of the technique were shown to be useful in capturing the unexpected impact of real-life variations of species on surfaces in the field. This is important work that continues to be funded by customers.

Single-Beam Femtosecond Multiplex CARS

This work illustrates the outstanding evolution of research aimed at using a collinear approach to coherent anti-Stokes Raman spectroscopy (CARS) for trace gas detection. These studies focused initially on pulse characterization but transitioned to the examination of mathematical methods and algorithms for extracting the desired spectral signal from broadband background spectra. The principal investigator was able to demonstrate strong signal-to-noise ratio improvements that substantially enhance the efficacy of the CARS approach.

Photon Trap for Infrared Detection

This project involved the expanded modeling and experiments on the microresonator enhancement presented 2 years earlier. The work showed that very small variations in microresonator dimensional control had strong impact on both the peak efficiencies and the bandwidth of the enhancement. The results were encouraging, indicating that design regimes existed where very high efficiency could be supported over a band that was easily large enough for many Army applications.

This important advance may not be receiving sufficient resources to move quickly to highly optimized commercial technology. Also, ARL's studies of the dynamic behavior of materials are likely to be advanced by improvements in infrared detection at modest elevated temperatures.

Ultrafast Spectroscopic Noninvasive Probe of Vertical Carrier Transport in Heterostructure Devices

This work involved pump-probe studies of ultrafast carrier dynamics and charge transport in heterostructures, with the ability to interrogate charge-generated terahertz field profiles in materials prepared for device structures. This research represents a valuable investment in advanced characterization, and the quality of both the topics and investigators is excellent. In addition to being of immediate value to materials and device researchers, the projects are conducive to quality papers and conference presentations of broad interest to the technical community.

Tunable Solid-State Quantum Memory Using Rare-Earth-Ion-Doped Crystal, Nd^{3+}:GaN

This project involved high-risk, early work aimed at using GaN as a host material for the neodymium ion (Nd^{3+}) in quantum memory research. The objective of this project is to perform photon echo experiments to provide an estimate of the memory storage time and capacity in cryogenically cooled Nd^{3+}:GaN crystals. The plan is to ultimately fabricate GaN polar heterostructures from which to design a quantum memory device with multimode capacity. Although this project is in its early stages, this work makes strategic use of ARL's strong GaN materials and molecular beam epitaxy (MBE) growth capabilities to gain a competitive position in a field that is drawing worldwide attention. Moreover, the ARL team has a strong track record of published contributions in this field.

Opportunities and Challenges

The consistent development and extension of modeling to broader sets of problems and applications is an opportunity area. One prototype project is short-wavelength IR device modeling and optimization. This research illustrates ARL's expanded efforts to provide critical modeling support in areas where there is high investment in underlying materials and device technologies.

The important software tool set coming from this research is not only essential for designers, but it may also provide critically sensitive parameters that could be used in process control for commercial partners and suppliers of imaging solutions to the Army, which necessitates engaging with the manufacturers. The project's principal investigator has started this engagement.

HIGH STRAIN RATE AND BALLISTICS MATERIALS

The overall impression of the materials in extreme dynamic environments program and the high strain rate and ballistics materials research at ARL was positive. The projects showed an excellent degree of integration between materials science fundamentals and applications, combining simulations and experiments aimed at developing structure-property correlations with advanced processing and fabrication approaches.

ARL is also establishing itself as a world leader by building novel capabilities, including, for example, extensive facilities for metals, polymers, and composites processing. The miniaturized Hopkinson bar and multiscale, rate-dependent mechanical testing equipment along with microscale sample preparation set-up for investigating polymers, metals, ceramics, fibers, and threads, are unique facilities. The commitment by ARL to take advantage of the dynamic sector facilities at the Advanced Photon Source is noteworthy. In situ measurements performed using these facilities will provide the needed fundamental knowledge for developing and validating computational models for improved understanding of high-strain-rate effects. Throughout the high-strain-rate and ballistic materials efforts, there is substantial growth in the use of computation and modeling and its integration with experimentation. Continued advances in this area are needed, adding, wherever possible, physics-based analysis.

Accomplishments and Advances

Materials in Extreme Dynamic Environments

This program is aimed at establishing the capabilities to design materials for use in specific extreme dynamic environments. It has considerable synergy with ARL's mission and is having a significant impact on internal and external collaborations, with about 54 faculty and senior scientists from various universities collaborating with about 35 ARL scientists. The collaboration has obvious advantages and is producing good fundamental research as evidenced by about 94 jointly authored publications. It is also promoting internal collaboration between different subfields and disciplines with expertise in computations, experiments, and manufacturing—pushing the envelope further in terms of the research conducted. The choice of simple metallic (Mg), ceramic (boron carbide, B_4C), polymers (ultrahigh-molecular-weight polyethylene [UHMWPE]), and composites (S2 glass/epoxy) is ideal for a materials by design approach. The "see it, understand it, control it, design it" paradigm is healthy and keeps the researchers focused on this holistic approach to the research. Additionally, interactions with the Defence Science and Technology Laboratory in the United Kingdom and the establishment of the Mach conference[2] are laudable efforts. The level of the research at ARL is benefitting significantly from these interactions. ARL is providing the necessary leadership to ensure this collaboration.

Methodology for Scale-Bridging in Multiscale Modeling of Materials

This program concerns the development of practical multiscale modeling strategies for applications relevant to the simulations of structural and energetic materials. The emphasis is on the use of small-scale models within elements, in effect, as constitutive models. The details of the small-scale models are invisible to the continuum-scale finite element code. The use of surrogate subscale models is a key innovation. This is a good idea because it allows subscale results to be recycled to future time steps, avoiding the need to repeat small-scale calculations. The effectiveness of the method results from the fact that the subscale model is applied to only a small fraction of continuum elements at any given time. This results in a significant reduction in computational time. Much of the development has focused on the computational infrastructure to make this strategy effective. The infrastructure adaptively assigns processors to the calculations in the two levels. The software tools make efficient use of many processors in a medium-sized application.

[2] The Mach Conference showcases the state of the art of multiscale research in materials, with an emphasis on advancing the fundamental science and engineering of materials and structures in extreme environments.

Polymer Modeling Research

This program is an excellent example of coordination between various topics aimed at microscale experimental investigation of UHMWPE fibers, including processing of fibers and films, and their effects on ballistic performance. The modeling examines the three-dimensional network of polymer chains as a double network, at an atomistic level, allowing for development of simulations of mechanisms associated with polyethylene. One of the keys to understanding this system has been the development of the reactive potentials used for modeling UHMWPE. The use of quantum mechanical data to develop this potential has been shown to be fast and accurate.

Graphene and Two-Dimensional Polymers

This program is enabling the design and exploration of properties of new materials—for example, different variants of graphene, or graphylene—that may lead to better armor. The possibility of developing a two-dimensional (2D) structure from polyethylene to increase ballistic performance is indeed an out-of-the-box thinking. Graphylene, an example of a proposed 2D polymer, consists of carbon rings connected by polyethylene chains. Density functional theory (DFT) and MD modeling of this polymer suggests that the carbon rings make it compliant and ductile while also providing the stiffness, strength, and fracture toughness needed for increased ballistic performance, in particular for body armor applications. This project, which is also leveraging an Army Research Office-funded multidisciplinary university research initiative program on 2D polymer synthesis, is an example of successfully using computational tools to design and develop new materials.

Grain Boundary Modeling and Simulation for Lightweight Protective Materials

This program included a fundamental study on B_4C that addressed the important issues at the root of its poor ballistic performance. The computational work, which can also be extended to other ceramic systems, is leading to an improved understanding of the amorphization process that occurs under the combined effects of pressure and shear. The potentials developed by Goddard and others were compared, through the virtual diffraction patterns generated, to experimental ones by Anselmi-Tamburini. The overall goal was to investigate the improvement of toughness mediated by engineered grain boundaries. The work represents an excellent fundamental effort.

Synthesis and Multiscale Rate-Dependent Response of Fibers as a Function of Microstructure

This project represents an impressive array of experiments used to establish the fundamental deformation mechanisms in polyethylene (PE) and UHMWPE. The strain-rate range investigated up to approximately 10^6 s^{-1} is impressive and is being enabled with the use of a miniature tensile Hopkinson bar. The gripping of the specimens is especially crucial and seems to have been successfully accomplished. Diagnostics, including digital image correlation (DIC) and X-ray measurement techniques, are revealing details about the internal process of failure within fibers that are apparently a first in the polymer fiber community. Future small-angle X-ray scattering characterization is planned, and fibril-level testing will be conducted to establish the effects of the spatial scale. The tensile strength of UHMWPE fibers is very remarkable: 3 GPa. The strain-rate sensitivity is surprisingly low and attributed to viscoelastic effects. The project is providing valuable data about the high-rate deformation and failure of polymer fibers that will be of lasting value to the armor material design community. The project is well aligned with other projects on polymer armor materials. Multiaxial loading is important in armor materials because the shear strength can depend on other stress components and the hydrostatic pressure. The future work involving atomic force microscopy (AFM) in real time is well thought of and at the frontier of knowledge.

The project is testing micron sized fibers to assess multi-axial stress states. Handling such small samples is no trivial task, leave alone getting high quality data on the material. An impressive array of experiments is being used to establish the fundamental deformation mechanisms in UHMWPE. Multi-axial loading is being initiated, with compression on the fibers superimposed along with tensile loading. The discussion of setting up a fiber and film facility, with future plans of adding additional diagnostics such as dynamic X-ray system is indeed exciting and fascinating.

Grain Boundaries and Interfaces

This project addresses the inelastic deformation due to contact loading of B_4C. Both material and model development efforts were supported by transmission electron microscopy (TEM) examination of sections subjected to inelastic deformation, which consisted of a Knoop hardness indentation. Increase in the load resulted in increasing planar defects, amorphous bands, and micro- and macro-cracking. Most amorphous bands followed the maximum shear stress trajectories.

Modeling and Performance of UHMWPE

This project is intended to develop an understanding of the connection between processing of composites with UHMWPE fibers and their ballistic performance. The researchers identified inter-laminar shear strength as a key material property and developed a simple but effective test to measure this property in a laminate. DIC diagnostics are used to visualize the deformation and crack growth at the interface. Finite element modeling is used to help understand the progression of failure. The main process variable being studied is fabrication pressure. Higher pressure decreases the inter-laminar shear strength. The researchers measured changes in the morphology of the ply interfaces that apparently help to explain this effect.

Multiscale Material Characterization, Modeling, and Experimental Methods for Fibers and Fabrics in Soft Armor Protection Systems

This project is developing and applying testing techniques to observe the microscale processes that occur in the interior of UHMWPE fibers during mechanical loading. ARL researchers have developed a novel sample preparation technique that helps to reveal the mechanics of fibrils within a fiber. Microscopic imaging techniques, including AFM, show the evolution of structures and defects as a fiber is strained. Force-displacement measurements are correlated with these microscale events. The techniques and data being developed in this project significantly advance the state of knowledge of the behavior of fibers beyond what is available in the literature. This project is providing unique insights into the materials science underlying the failure process in fibers. The data collected and the new techniques developed will be of great value in the improvement of fiber materials and processes. The data will also be useful to computational model developers.

Depleted Uranium Replacement

This project builds on the on-going search for new penetrator materials. Currently, tungsten-based alloys are used, but the performance is inferior to that of depleted uranium (DU). One important feature of DU penetrators is the formation of adiabatic shear bands, leading to a reduction in the diameter of the hole and increased penetration depth. The current research on nanocrystalline tungsten powders and their consolidation by pressure-less sintering seems to be producing materials with desirable properties with the ability to shear localize.

Optimized Tungsten Carbide Materials for Improved Lethality

This project uses nanocrystalline iron to bind tungsten carbide (WC) powders that show excellent properties. This binder replacement is motivated by the toxicity of the current cobalt binders. Pressure-less sintering is used to produce the necessary homogeneous microstructure with the desired hardness and toughness properties. The promising feature of this project is the repeatability of the process, which is the first step necessary for further studies prior to transition. There is a close collaboration with the lethality group and the U.S. Army Armament Research, Development and Engineering Center (ARDEC) for further evaluations. The fracture toughness of the WC-nanocrystalline iron is equivalent to that of the conventional WC: about 12 $MPa.m^{1/2}$.

Exploiting Oxide Dispersion Strengthening in Ferritic Alloys for Lethality Applications

This project is using the scale-up capabilities available at ARL for high-energy ball milling of powders and equal channel angular pressing processing to synthesize and fabricate oxide-dispersion strengthened ferritic alloys. The bulk material fabricated with a microstructure consisting of nano- to microscale grains with larger-sized intermetallic precipitates and zirconium oxide dispersed particles, demonstrates substantially high room-temperature compressive strengths of the order of 1.2 GPa for those consolidated at 700°C, and 2.4 GPa for those at 1000°C. Although the compressive strength at elevated temperature is lower than that at room temperature, the work illustrates the control of interstitial elements picked up during powder processing.

Superhard Diamond-Diamond Composites Made via Laser Chemical Vapor Infiltration

This project combines selective area laser sintering and localized laser chemical vapor deposition with motion control and real-time temperature feedback loop to fabricate bulk and complex shapes from diamond-diamond powder composites. The technique uses conventional additive methods to first create a preform starting with diamond powders, and next follows that with chemical vapor deposition of diamond coatings on the particles in the preform to grow and form the fully dense diamond-diamond composite. The approach combines the benefit of both processes, which individually are unable to produce diamond powder compacts in bulk and complex forms. The approach also provides the versatility to employ various ceramics to create the matrix (binder) phase for the diamond preforms. The acquired components for building the reactor for the deposition system have been assembled, and the control program is being developed. Once the system is built and the control program is optimized, it will be used to identify the sintering conditions for fabrication of the diamond preforms and the desired deposition chemistries to achieve the growth conditions needed to make bulk composites with different reinforcement and matrix phases. This is a good example of a project in which an ARL postdoctoral researcher is building on the technology developed as part of their Ph.D. dissertation and extending it to a new capability at ARL for making diamond-based composites for potential armor applications.

Mechanics and Performance of Unidirectional Film Laminates for Ballistic Protection

This project is investigating the effect of processing conditions for laminates made of UHMWPE films. Film laminates behave differently from the more conventional UHMWPE fiber-reinforced laminates. The film laminates do not need a separate matrix material like epoxy. However, their mechanical properties are sensitive to processing conditions, especially temperature. The film plies bind to each other just below the melt temperature. The properties of the film laminates are unknown, compared to the fiber laminates. However, it has been discovered that under suitable processing history, the film laminates have excellent mechanical strength, even without the inter-laminar matrix material. The project is also seeking out lower-cost solutions to fabrication. It is an important project to the ARL mission of developing effective and lightweight armor materials. Important results are being obtained that will aid the application of film composites for land forces.

Modeling and Characterization of Two-Dimensional Polymer Ultra-Membranes

This project is using atomistic modeling to demonstrate the possibility of a new class of a graphene/polyethylene 2D material, termed graphylene or GrE-2, with enhanced fracture toughness while maintaining twice the stiffness and nine times the strength of Kevlar. The MD simulations considering two-carbon long ethylene chains connecting the benzyne rings in the polymer framework, illustrate that the energy release rate for crack propagation in graphelyne is twice that for graphene due to effects of dissipative processes such as delocalized failure and crack branching. The simulations were extended to predict the response of ensembles of discrete platelets of 2D materials and to demonstrate superior mechanical performance through careful design of inter-layer interactions. This is the first such study predicting the design and mechanical behavior of 2D materials aimed at guiding the synthesis of novel polymers. These efforts have yet to be extended to the prediction of the performance under high-rate impact conditions relevant for applications in body armor or to the synthesis of such a material.

Peridynamics Modeling of Projectile Impact and Penetration in Vehicular Glass

This project is a novel computational simulation method that is intended to predict the impact resistance of glass armor materials. The peridynamic method is compatible with the physical nature of the problem because it allows for discontinuous deformations, since it does not rely on differential equations. The peridynamic equations do not require smoothness of the displacement field. The method therefore has inherent advantages over traditional approaches to fracture modeling that rely on finite elements to approximate the partial differential equations of the standard theory of solid mechanics. The researchers have developed a peridynamic solver that couples with a finite difference solver, allowing the peridynamic portion of the mesh to adaptively follow the growing crack tips. This innovation allows for greater computational efficiency and reduced wave dispersion in the coupled model. The project team is acquiring an ISRA optical instrument that can measure flaw distributions in glass samples prior to impact. This will allow measurements of flaw distributions to be initialized into the computational model, thus providing more meaningful validation of fracture and fragmentation than was previously possible. The principal investigator (PI) collaborates with other staff members at ARL who fabricate glass with specific compositions, permitting the modeling effort to be coordinated with materials processing. The modeling technique and validation data being developed in this project will provide valuable capabilities that are unique to ARL and directly align with ARL's mission to design and improve armor materials. ARL is establishing itself as a widely respected center of research on the peridynamic theory and computational methods. The project also contributes to ARL's leadership status in the field of transparent armor impact mechanics and materials science.

Understanding the Role of Glass Composition on Properties and Performance

There were several impressive posters reporting on new measurement techniques and diagnostics capabilities. For example, the exciting and novel development of a miniaturized Hopkinson bar (also known as a Kolsky bar), is used for characterization of polymers, metals, ceramics, fibers, threads, and fibrils, at high strain rates. This will help with the development of new models or validation of models at strain rates higher than those that are currently accessible. The development is novel and a breakthrough in the ability to understand materials at the micro scale. Coupled with time-resolved diagnostics such as DIC, dynamic X-ray technology, framing cameras, and so on, this capability has the potential of making ARL a leading research institution in investigating material behavior and properties measurements at high strain rates. The projects that highlight these capabilities are described below.

Advanced Experimental Techniques

This project focuses on sample preparation and fabrication processes for characterizing materials at high-strain-rate loading. Using lasers, mini samples with dimensions of about 500 μm can be made. This technique was developed at University of California, Santa Barbara. Large numbers of specimens can be produced for quasi-static and dynamic testing.

High-Strain-Rate Deformation Mechanisms of Polymers

This project uses new capability to characterize polymers such as polycarbonate and polymethyl methacrylate over a broad range of strain rates as accessed by the mini Hopkinson bar. A range of capabilities was used to establish the response of bulk polymers (primarily polycarbonate and acrylic) over a broad dynamic strain rate regime. Infrared detectors measure the increase in temperature due to deformation, which is especially important in polymers where the glass transition temperature is not far from room temperature.

Rate Dependent Mechanical Response of Polymer Networks

This project is exploring another application of the mini Hopkinson bar for testing of epoxy-based polymers. It involves an integrated modeling and simulation approach to identify possible chain mechanisms to explain the improved strength performance observed in high-strain-rate measurements on epoxy.

Opportunities and Challenges

Methodology for Scale-Bridging in Multiscale Modeling of Materials

In its present state, the multiscale computational method apparently has been applied only to subscale models that essentially provide an equation of state to the continuum finite element model. While the equation-of-state work represents a significant achievement and promise for this technique, the incorporation of other aspects of material response, such as evaluation of the full stress tensor, which can be challenging, has not yet been addressed. Realistic subscale models for materials involve the evolution of defects that lead to phenomena such as work hardening, thermal softening, shear localization, dynamic fracture and spallation, and so on, which remain to be incorporated. The example provided in a presentation to the panel, a Taylor test, is an oversimplified case. The incorporation of physics is an essential part of predictive codes and cannot be ignored. In the case of explosives and propellants, chemical reactions, deflagration, and detonation have to be incorporated. Close work with Betsy Rice, a globally recognized leader in this field, is highly recommended. The stated goal of connecting quantum mechanics, MD, dissipative particle dynamics (DPD), and continuum models will require the ARL team to investigate multiscale strategies and go beyond the scale-coupling focus and software aspects of multiscale computational modeling. Furthermore, although these aspects present significant challenges that the ARL team is addressing very well, the physical models underlying the subscale computations are worthy of careful research and cannot be treated as a black box. For example, it cannot be assumed that data estimation from DFT subscale models will work the same as with MD or DPD. Collaborations with researchers at the California Institute of Technology (e.g., with Michael Ortiz and William A. Goddard) and other institutions, as well as at the Department of Energy (DOE) laboratories, may be worthwhile to seek paths for the incorporation of the physics in the subscale models and their implementation into the multiscale infrastructure.

Polymer Modeling Research

The methods used in this program can be applied to link continuum models to better understand mechanisms in more bridged structures. Understanding of the mechanisms of these materials and the interface interactions has the potential to lead to improved polymeric armor with better ballistic performance.

Graphene and 2D Polymers

The goal of computational materials design is still largely unrealized in the larger multiscale modeling community. If the ARL group can succeed, the payoff would be very high. Currently, materials such as Kevlar and boron carbide are used for body armor applications. The use of materials such as graphylene would have the potential of a significant reduction in body-armor weight. The ARL effort in this area appears promising, but the activity is largely academic if the material modeled cannot be synthesized. This program needs to include a strong synthesis component.

Composite Materials in High-Strain-Rate Environments

This program is intended to evaluate a set of continuum-level material properties for composites on the basis of computational modeling at smaller-length scales, with emphasis on the use of modeling to optimize the materials composition. The prediction of the composite response from the properties of its building blocks (constituent materials and interfaces) is an exceptionally difficult challenge that has frustrated researchers over the years, partially because of the multiplicity of length and time scales whose mutual interaction affects the progression of material failure. The ARL team has access to new computational tools that help to address the multiscale aspects of the challenge. The divergence in time scales from picoseconds up to milliseconds presents major difficulties. On the other hand, the objective of understanding and optimizing the properties of individual constituent materials and interfaces appears more tractable. For example, the use of network models for epoxy to improve its properties seems promising. Similarly, the use of MD to understand and improve polymer fiber performance is a reasonable objective. The use of mesoscale models to reproduce and predict the failure modes in woven composites is a sensible plan. The progress composite damage model method for homogenizing and scaling up mesoscale damage simulations also makes sense as part of the overall analysis. The traction versus displacement data that were shown for epoxy-glass interfaces appear to reveal that such interfaces fail under tension at around 5 GPa, which seems high and far exceeds the strength of any epoxy. In summary, the prediction of a full set of LS-DYNA MAT162[3] parameters for a real composite solely on the basis of MD and mesoscale computations is perhaps not within reach. The challenges in doing so are enormous, and the investigators need to be objective about the prospects for really accomplishing this. Nevertheless, the computational mechanics community learns a lot by trying to address such challenges. Furthermore, the individual component models for constituent materials and interfaces have value in themselves and may lead to improvements in real materials by providing fundamental understanding. It is good that the modeling team has experience with the practical aspects of composite materials science and processing. It would be helpful to include team members who can synthesize the new materials to confirm the results of the computations in the laboratory, if this is not already being done.

Physical Response of Boron Carbide Under Extreme Conditions

This project involves connecting the quantum-level response to macroscale impact, combining characterization approaches and penetration studies to demonstrate improved performance. B_4C is a material of interest to the Army because it is used in body armor, due to its relative low density and high strength. However, unlike other ceramics, there are two distinct features that are observed experimentally. One is the loss of strength of the material under uniaxial-strain shock or dynamic loading conditions, and the other is the observation of localized amorphous phase observed in recovered specimens under bullet or ballistic impact. There has been extensive discussion of strength loss of B_4C leading to a lower than anticipated ballistic response of the material. There has been speculation that these two events are related. This has led to a detailed study identifying other physical mechanisms for the observed strength loss. Mechanisms such as amorphization, friction, localized melting, oxidation, and others that were proposed to the panel as reasons leading to strength loss were not conclusive. The technical aspects presented in trying to identify mechanisms for strength loss were quite detailed but not successful. What leads to amorphization is still not clear, nor is it confirmed that it is the primary mechanism responsible for strength loss. B_4C is in extensive use and, as a practical matter, it is not clear if significant timely and substantial improvement in performance can be made. However, as a research exercise, it may be necessary to tie loose ends to make an impact on the design and performance of similar future ceramics when deployed.

Grain-Boundary Modeling and Simulation for Lightweight Protective Materials

The researchers work collaboratively with experimentalists in an effort to verify their calculations and thereby generate a better understanding of the effects observed in B_4C and other ceramics.

---

[3] LS-DYNA is a finite element code and MAT162 is a material model for use in LS-DYNA that may be used to simulate the onset and progression of damage.

Grain Boundaries and Interfaces

More work is needed to understand the morphology that has been observed to-date and to relate the results to the dynamics of the system. The methodology could also be used to examine other systems. While the TEM examination of indented samples is being used, one of the major challenges will be the need for operators knowledgeable with the techniques and the scientific information that can be extracted. Sample preparation and interrogation require a level of understanding unlike other techniques used for characterization.

Atomistic and Mesoscale Modeling of Grain Boundaries

This project involves DFT and MD work aimed at explaining how grain boundaries can be engineered to increase the toughness of B_4C. The role of intergranular films is still not being evaluated, but it is a goal. The role of amorphization of B_4C and its relationship with failure is being evaluated. Failure is another instance of critical phenomena related to approaching a percolation threshold. This is a fundamental research project, and applications are still far away. Nevertheless, the fundamental knowledge gained justifies this effort, especially in view of the fact that the methodology can be applied to other ceramic systems, such as boron oxide.

Modeling and Performance of UHMWPE

Surprisingly, increasing the inter-laminar shear strength of materials has a detrimental effect on ballistic performance. It would be interesting to apply computational modeling to understand why this trend is observed. The project will help to standardize the processing of UHMWPE materials by identifying an optimal set of conditions for ballistic resistance. This in turn will help improve the uniformity and quality of armor materials.

Depleted Uranium (DU) Replacement

Nanocrystalline tungsten still lacks tensile ductility, but processing improvements could lead to improved performance. The research being performed will identify the causes of strain localization and determine critical mechanisms of plastic flow at various strain rates. The project is challenging and is the basis of the doctoral dissertation of an ARL researcher at Johns Hopkins University under the advice of Kevin Hemker. This is a good example of an internal ARL project being used to develop the professional growth and career for existing early-career, bright, and talented staff. This project needs to incorporate wires (W or steel) to increase the tensile strength of the penetrators and their survival during launch.

Optimized Tungsten Carbide Materials for Improved Lethality

This is a perfect example of how ARL researchers are finding applications for powder metallurgy in which a fine-grained, pressure-less sintering process appears to have a high payoff. Current scale-up efforts and ballistic testing will determine whether this new material can be incorporated into weapons.

Exploiting Oxide Dispersion Strengthening in Ferritic Alloys for Lethality Applications

This work demonstrates the potential scale-up possible with the capabilities available at ARL guided by in-depth processing-structure-property correlation studies. Developing dispersion-strengthening analytical models would be helpful in determining the role of dispersoid particles on high-strain-rate properties, as oxide dispersion-strengthened materials typically show sustained strengthening at high temperatures due to the stability of the oxide dispersoids.

Mechanics and Performance of Unidirectional Film Laminates for Ballistic Protection

After further progress is made on the optimal processing conditions, it might be helpful to include a continuum-level modeling component to the project. This might give insight into the thermal and mechanical history in the interior of a laminate during processing and help reveal hard-to-measure effects such as thermal residual stress.

Understanding the Role of Glass Composition on Properties and Performance

This project is an experimental investigation to develop improved glasses as a transparent armor for lighter, durable protection systems. Commercial soda lime and borosilicate glasses currently deployed are not optimum armor materials, and without an identified consumer need for improved glasses, it falls to ARL to develop these materials using an in-house glass processing facility. The focus is to understand the role of Na-B-Al-Si in the casting of glass. The goal is to tie in and identify the role of composition to ballistic performance. As a part of this effort, the researcher is using finite difference peridynamics to model penetration events. It is not clear, however, if there is a modeling effort that correlates the composition of added impurities that would result in identifying the necessary dynamic properties that lead to optimum performance. It would be a benefit to the glass manufacturing program if such a modeling program existed.

STRUCTURAL MATERIALS

Within the structural materials portfolio, there are projects tackling tough, challenging problems in collaboration with ARDEC and other Army users to understand the importance and impact of the work to the success of the mission.

The research approach of coupling modeling with experimentation was evident across many of the efforts reviewed. Of note are programs designed not only to support specific and narrowly focused materials development efforts but also those that will produce tools that may be used more broadly. An example of one such effort is a program directed toward grain boundary modeling of ceramics for light-weight protective materials. A suite of tools is being developed to permit simulation of grain boundary structure and properties under high-rate loading conditions. Although these tools are being used to investigate grain boundary structure and properties of boron-based lightweight ceramics, these same tools will be applicable to the study of grain boundary interfacial relationships across all ceramic materials.

The potential wide applicability for these modeling and simulation techniques has provided the motivation for sparking university collaboration and cooperative research and development agreements (CRADAs) with industry. It is projects like this one that will provide ARL with the capabilities it needs to respond rapidly to future threats.

Accomplishments and Advances

Grain Boundary Modeling and Simulation for Lightweight Protective Materials

This program is focused on simulation of grain boundary structure and properties of boron-based lightweight ceramics—B_6O and B_4C. The key technical challenge is to discover and exploit grain boundary interfacial relationships at the nanoscale. This can inform experimental efforts to manipulate interfacial properties to optimize lightweight protective armor materials. A suite of tools is being developed to permit simulation of grain boundary structure and properties under high-rate loading conditions.

The research is complicated by the triclinic crystal structure and the existence of many boron-based ceramic polytopes. The tools developed include code to generate initial (unrelaxed) structures based on the 5 degrees of freedom of a grain boundary and simulation of diffraction patterns of polycrystalline material. Simulations of grain boundary relaxed structures and properties are planned after interatomic potentials for boron-based ceramics are tested and refined. Tools are also under development to derive grain boundary mobility and fracture strength once reliable potentials are available.

This is an excellent beginning to model and ultimately improve the properties of grain boundaries and hence polycrystalline boron-based ceramics. These tools may be extended to other crystalline material systems. The effort has collaborations with a large number of universities and is developing CRADAs with various industrial companies.

Nanostructured Metallic Materials

An important nearer-term potential application is a shaped charge liner strong enough to contain a hyper-velocity jet and sustain severe strain rates of 10^6 to 10^7 at 200 GPa pressure. It is known that a shaped-charge jet length improves at smaller grain sizes, but no data exists for performance below 10 micron grain size; it is possible that nano-grained copper may provide substantial improvement. It is hypothesized that the Cu-Be materials can be replaced by nanocrystalline Cu-Ta using a nonequilibrium processing method if the material can be effectively produced in bulk form; it is important to address this challenge. Small batches of Cu-10Ta have demonstrated high grain stability, since the Ta particles do not dissolve in the copper and effectively control grain growth. In addition, the material has high ductility, low creep rate, and exhibits superplastic behavior, making it a potential candidate material for a shaped charge liner.

This program is in its third year and is meeting the objectives. Future work will include experimentation with a Ni-Y system for higher-temperature applications.

Next Generation Lightweight Armor Ceramics

This is a follow on project to previous work on ceramic armor materials. Previous work resulted in transition of boron ceramic armor from companies such as CoorsTek. The objective for this new project is to develop new ceramics and geometric structures that deliver improved armor protection against small-arms fire with less weight. Desired weight reduction is on the order of 20 to 30 percent, and it is unlikely that existing standard boron or silicon carbide ceramics will achieve this objective. The hypothesis of the project is to use new, very hard ceramics. Initially, B_6O and $AlMgB_{14}$ are being studied. Both of these ceramics have better hardness properties than existing boron and silicon carbide ceramics. In order to use these new materials, the project has been divided into three tasks:

1. *Scalable synthesis and processing of the materials.* Powders for many of the desired ceramics are not available, prohibitively expensive, or of questionable quality. Sintering of ARL synthesized powder is being studied to assess sintering properties that lead to economic sintering of test coupons. These coupons are being ballistic tested at the subscale.
2. *Characterization of the novel heterogeneous microstructures that occur in these ceramics.* Understanding the properties of these unique heterogeneous microstructures is not well established. Microstructure models and resulting properties are being correlated to ballistic testing and damage mitigation.
3. *Development of processing science and exploitation of heterogeneous microstructure for improved armor performance.* This task is just starting. Exploration of previous related theory, process development, and modeling is under way. Methods for heterogeneous layering and feature creation for better protection is also being studied.

The project is aligned with Army needs, and team members are coordinating with Program Executive Office Soldier and the industrial community. The project appears to be on track with superhard ceramic powders being produced at ARL in the quantity required for sintering experiments. Sintering samples and small-scale plates have been produced. These samples are being examined for microstructure and unique microstructures have been identified.

Grain/Interface Boundaries in Boron Carbide Due to Contact Loading

This project addresses the formation of regions in B_4C that are interpreted as amorphous bands. The amorphous bands are generated with Knoop microhardness indents and are analyzed in cross-sectional TEM analysis. The main project objective is to assist in interpreting the indentation size effect. This effect describes the rapid increase in hardness with decreasing indentation load. In ceramic materials, this effect is attributed to the formation of cracks and bands such as the presumably amorphous bands observed in the B_4C indentation. The project has yielded TEM images that are used to rationalize earlier results and parallel efforts by other ARL personnel.

The TEM images show that microcracks develop from the linear bands. However, the link to the overarching objective of explaining the indentation size effect is still in question.

Three-Dimensional Through-Thickness Reinforcement Composite Armor Materials for Advanced Vehicle Protection

This is a mature project in its sixth year. The effort involves development of a complex model at the material level that is able to predict the fracture and energy dissipation of a very complex arrangement of materials, including ceramics and woven fibers. By looking at experimental impact data, the PI has quantified the crack initiation and propagation in the ceramic as well as at the multiple material interfaces. The model has been used to design armor that performs as well as conventional steel armor at half the weight. A cost/material/fabrication model has been developed that can be used to select an optimum design at the system level. There is one application in production and a number of others under evaluation.

The project team has excellent knowledge of the technical challenges involved, and even though the cost model was not developed in the most scientific way, all of the relevant issues have been addressed in an appropriate manner. The results met the project objectives.

Multi-Functional Multi-Material Topology Optimization for Lightweight Unmanned Systems

This is a computationally intense design-oriented effort with the potential to disrupt maneuverability and protection capabilities in unmanned systems. This agility goal is motivated by the Army's identification of remote and autonomous systems (RAS) as a vital effort to maintain operational advantage and overmatch on the battlefield. In order to lighten the burden of the soldier, augment protection, increase mobility, and enable joint combined arms maneuvers, advanced topology optimization methodologies for the design of multifunctional, multimaterial components are formulated and implemented to enable and fully exploit unmanned systems.

The work has demonstrated the use of computational algorithms for designing some model actuators. A demonstration platform was developed by combining materials with different moduli to expand the design window for actuator systems.

Defect Engineering of Nanocrystalline Alloys from Powered Precursors

This project is intended to develop bulk, thermally stable nanocrystalline materials with tailored properties for future force and ballistic applications.

Cu-10T, an alloy powder, is prepared by ball milling and consolidated by equiangular extrusion, which produces a solid solution of Cu and Ta, despite being completely insoluble. Upon heating, TEM investigation showed nanometer-sized Ta particles that pin the fine 150 nm grained Cu structure. Heat treatment up to 1000°C showed very limited grain and particle coarsening. The Ta particles appear to have an amorphous core, but further research is necessary for confirmation. Atom probe tomography is planned to determine the extent of carbon and oxygen contamination of the interphase boundaries. Collaboration with the Mishin research group at George Mason University is providing theoretical support of the coarsening processes through MD simulations. The simulations and TEM observations show only Ta particle coarsening at grain boundaries.

This is excellent research that combines theory, experiments, and analysis. Work remains to understand mechanical properties at elevated temperatures.

High-Performance Powder Metallurgy Titanium

This project has the objective of producing low-cost (estimate $1 to $5 per pound for unalloyed powder) powder that can be consolidated into various Army applications by additive manufacturing or metal-injection molding processes with properties equivalent to those of wrought titanium. The effort employs a unique process of hydrogen sintering and phase-transformation process developed at the University of Utah. Variations of this process can use blended elemental powder, titanium sponge, fines, and turnings to make alloyed and unalloyed powder, achieving densities of 99 percent as sintered, and approach 100 percent density with gas isostatic forging technology. The material produced through this process may be strong and tough enough for armor applications and components for a machine gun, among others. It has not been demonstrated for aerospace or high-temperature applications.

This is a new fiscal year (FY) 2016 effort with great potential. If this effort is able to demonstrate feasibility, further development of the process will be required for scale up. Work so far meets the program objective, and the path forward has been delineated. Work scope includes identifying the most promising Army applications so that demonstration parts can be produced. This approach will allow the Army to make an informed decision regarding the scale-up of this technology.

Corrosion Science

This research effort is intended to develop an understanding of fundamental corrosion mechanisms of metallic materials used by the Army and then apply this knowledge to develop suitable corrosion-mitigation strategies. It is currently focusing on magnesium alloys corrosion in aqueous environments—including understanding of corrosion initiation sites, influence of alloy chemistry, trace element effects, and so on. The research employs state-of-the-art instruments, such as an electrochemical atomic force microscope.

The effort is in its fourth year and has produced a predictive atomic level model (based on density function theory) of magnesium corrosion. Various mitigation strategies suggested by the model are being tested.

Extreme Mechanical Response of Electro-Chemical Fabricated Nano-Crystalline Metals

This is an early research effort focused on grain refinement to the nanometer scale through a systematic approach of electrochemical fabrication, unique high-rate testing, and mechanistic crystal plasticity modeling. This research is an excellent demonstration of the approach to tightly couple modeling and experimental work. This project will develop an understanding of the mechanical response of grain refinement in three common crystal structures at high-deformation rates.

The success in developing small-scale fabrication and testing capabilities will chart new territory in the understanding of grain refinement and strain-rate design space. This project leverages work from the advanced experimental techniques project and greatly reduces the laser setup time for sample preparation from days to hours. While miniaturization of the split Hopkinson pressure bar test for very high strain-rate testing is not new, the testing of strain rates up to $10^6\ s^{-1}$ is novel. The microscale sample preparation and miniaturization of the Hopkinson test demonstrates the resourcefulness of the researchers and collaboration across projects. This work is foundational, and the researchers are encouraged to continue to look at synergies in other research areas, such as the enhancement of corrosion-resistant materials.

Functionally Graded Protective Coatings by Materials by Design

This project addresses an urgent Army need. Sand is readily ingested into jet engines where it melts in the hot section of the engine and adheres to component surfaces, rapidly degrading material performance. Large particles can be removed by various techniques, but small particles remain.

Superalloy components have ceramic coatings that act as thermal barriers. Modifying the coating (either the entire coating or a surface layer) could change the ability of the molten sand to physically penetrate the coating, forestalling degradation. Another possibility is that some kind of mechanical cycling could fracture and force off the sand layer after it re-solidifies.

The project includes spray synthesis of different ceramic coatings, followed by characterization and testing to simulate field conditions. The project is currently focusing on developing a basic understanding of the interaction of the molten silica with the surface of the coating. Incomplete information is available on basic properties of proprietary coatings. The project is at a relatively early stage; for example, no work has yet been done on graded coatings. Publications show excellent teaming on the project.

Interfacial Load Transfer in Ring-Opening Metathesis Polymer-Based Composites

These composites are considered candidate resins for ballistic resin-fiber composites. A drawback of this new class of polymers is the surface chemistry that hinders, if not prevents, bonding with fibers. The current research is an integrated modeling and experimentation project that addresses several aspects, including the interface engineering to achieve tailoring resin-fiber bond strength, the synthesis and scale-up of ring-opening metathesis polymer-based composites (ROMPs), and mechanical testing of the composites. Most research on resin-fiber composites is driven by aerospace applications and, therefore, does not consider high-strain-rate effects. A particular focus for the Army is therefore to develop capabilities to control the fiber-resin bond strength and to tailor the composites to varying ballistic conditions.

The approach to achieve ROMP-fiber interfaces suitable for high-strain-rate deformation on the experimental side is based on film and coupling agent chemistry experiments. Modeling efforts involve mostly MD simulations that are used to determine constitutive models for the interface that are then used for finite element method (FEM) simulations of the composite mechanical behavior. The predictions are evaluated with mechanical shear tests and sample geometry. A test was developed to facilitate comparison with the modeling predictions.

Opportunities and Challenges

Materials for Robotic Augmented Soldier Protection

This program is a systems engineering challenge that presents opportunities for the synergy of materials, design, and manufacturing. It addresses the need for an overall strategy for robotics in the battlefield and in particular for the role that materials will have in the different robotic missions. At this stage in the project, from a materials-driven perspective, the optimum suite of materials for different robotic missions has yet to be identified. Starting with an overall robotics strategy, a comprehensive materials science strategy could be developed. As an example, this project could be quite ambitious with the identification of materials needed by unmanned aerial vehicles to protect soldiers against various threats (ballistics, radio frequency [RF] bombardment, electronic warfare) and micro air vehicles to operate in different environments. It could include disruptive materials needed for energy absorption and utilization of incoming threats (ballistics, shrapnel, RF) for perimeter robots or soldier-assigned robots; materials needed for robots to survive extreme environments (e.g., fire, sand, corrosive, radioactive, volcanic ash) to perform dangerous missions, including in situ repair and maintenance in hazardous environments and rescue operations.

This project has the potential to be a strategic campaign with a focus on the triad—materials, design, and manufacturing—approach to multiscale, multimaterial, multidisciplinary, and multifunctional robotics strategy. Research and collaboration with the Defense Advanced Research Projects Agency's (DARPA's) robotics initiative and SPARC (the partnership for robotics in Europe) and understanding of the competitive landscape is needed if this project is to succeed.

Nanostructured Metallic Materials

This program is intended to develop stable bulk nanocrystalline metallic alloys from powder precursors for lethality applications. Nanostructured metallic materials can be unstable, making the retention of high strength and exotic properties challenging. Low thermal stability also constrains processing techniques, as well as service temperatures. Nanocrystalline metals usually have low ductility but high strength, making their application as a structural material difficult, requiring methods to control nano-grain size to achieve desired strength levels while retaining required ductility and fracture toughness properties.

Potential applications of these materials include advanced personal and vehicle armor, warheads and kinetic energy penetrators, and advanced structural alloys for weapon and surveillance platforms.

The program will continue to develop the foundational science of grain boundary solute engineering in nano-duplex structures, devise engineering methods to control deformation at the nanoscale, advance processing techniques capable of producing suitable bulk material, and pursue the design and development of components that utilize these nanostructured materials. It needs to be noted that related work for producing bulk nanocrystalline material is ongoing in Australia, the European Union, Russia, and China.

Grain/Interface Boundaries in Boron Carbide Due to Contact Loading

This project is part of a larger effort to understand the deformation behavior of B_4C under high-strain-rate conditions. The combination of Knoop hardness indentation and TEM cross-sectional analysis is used to reveal cracks, linear amorphous bands, and underlying microstructures. Additional work will be necessary to establish a relation between the linear bands, cracks, and the indentation size effect. Suggested additional experiments include annealing experiments that could be used to crystallize the amorphous bands and corroborate the amorphous nature of the bands. Instrumented indentation experiments could be used in conjunction with stress analysis to compare the experimental observations with predictions of maximum shear stress patterns in the cross-sectional images.

Multi-Functional Multi-Material Topology Optimization for Lightweight Unmanned Systems

Numerous potential applications have been identified, including structural radomes and structural capacitors, but much more work is needed before these applications are realized. A major gap in the research is to include microstructural sensitive properties into the topology optimization algorithms. Added expertise or collaboration with materials scientists would help.

Corrosion Science

As this effort moves forward, there will be a need to consider how the mitigation would be applied to Army hardware. For example, ion implantation does not perceptually change a surface, so detecting if an area has been adequately treated is challenging, and components with large surface areas may not be able to use this process. This research effort could consider applying a protective coating using cold spray methods.

Extreme Mechanical Response of Electro-Chemical Fabricated NanoCrystalline Metals

A follow-on project to investigate the differences in properties at the component level due to the influence of factors (such as geometrical features, volume, size, heat treatment, surface finish, and so on), compared to the specimen properties, would enhance the computational framework to validate the models of component-level structures and provide guidance on material-design-manufacturing choices. Researchers are encouraged to stay connected with the National Nanotechnology Initiative and the Institute for Soldier Nanotechnologies. To ensure that ARL continues to be at the forefront of this research, it is highly encouraged that researchers participate in nanotechnology-related conferences, such as the SPIE (the International Society for Optics and Photonics) nanoscience and engineering conference and the Materials Research Society conference on nanoscale materials.

Interfacial Load Transfer in Ring-Opening Metathesis Polymer-Based Composites

This effort began in FY2016 and ends in FY2018. The metrics for the project involve demonstrating the ability to synthesize and control interface bonding agents and to develop validated predictive MD and FEM models. So far, a bonding agent has been found that can be tailored to achieve different bonding strengths between fiber and resin. A conventional fiber pullout test could yield data that are more suitable for the project than the lap shear test that represents the composite as a whole, as opposed to the fiber-resin interface.

The project has clearly met a main objective—to identify a resin-fiber bonding agent that can be adjusted for different interface strengths. But at the same time it is not clear who and how this agent was identified: Did the input come from elsewhere? Was the bonding agent found empirically? Some of the references by the PI on the project seem to have been published in 2015, before the start of the program. The other project objectives have not yet been met.

NanoMicro Particle Integration for Composites

There is still much work to be done to meet the objective of developing computational algorithms to design game-changing, lightweight, multifunctional, multimaterial components produced via additive manufacturing. Numerous potential applications have been identified, including structural radomes and structural capacitors, but much more work is needed before these applications are realized. The project is not yet at the point of incorporating materials properties and processing into the topology optimization algorithms, and including microstructural sensitive properties in the algorithms is a major unrealized opportunity. Broadening the project scope to take advantage of materials design strategies and adding expertise or collaboration with materials scientists would help.

ELECTRONIC MATERIALS

Overall, the quality of the electronic materials applied research and development (R&D) efforts is outstanding with well-supported, long-range projects that are maturing and moving into manufacturing (MANTECH programs), balanced by new, advanced research initiatives.

The R&D efforts undertaken in response to the dramatic change in the battlefield environment—transformed from a relatively small number of large, well-supported divisions to smaller distributed groups—are impressive. This transformation challenges the Army to provide the infrastructure for supplies and, particularly, equipment repair. Many of the ARL projects reviewed are concerned with facilitating this transition. An exciting example of one of these efforts is the application of basic metallurgical science to provide field-deployable, individual custom repairs to significantly damaged vehicles—helicopters, armored vehicles, and so on.

Accomplishments and Advances

Vertical 2D-3D Semiconductor Heterostructures

This program captures one of ARL's more aggressive and admirable efforts to map basic research in advanced 2D electronic materials into Army needs. The underlying concept of this newly funded project is to harness vertical transport through 3D-2D-3D stacks, as opposed to more conventional in-plane strategies, to achieve bipolar-like transistors that may open up new performance regimes. Specifically, the project includes both experimental and theoretical efforts to optimize epitaxial growth, electronic interface functionality, and transport characteristics of 2D MoS_2 on bulk GaN with an eye toward applications for high-power, high-speed RF transistors.

This work is at an early stage, but the PI and the team have set out a promising multifaceted program that is well positioned to contribute to the basic understanding of the efficacy of fabricating and harnessing vertical transport in these structures. The initial epitaxy has illustrated islands of single-crystal MoS_2 films that have shown promising characteristics, including 20 times photoluminescence enhancement relative to exfoliated films. Experimental work has also included conductive AFM and Kelvin probe measurements to capture contact resistances, tip/MoS_2 Schottky barrier height, and work functions for the epitaxially prepared materials. While efforts to date have been all n-type, the team is moving to p-type GaN to enable evaluation of expected diode behavior.

Energy Efficient Soldier Radios

This program is directed to the critical need to improve power efficiencies in mobile, secure radio communications to relieve the battery weight burdens on the soldier, which can be as large as 60 percent of a soldier's load. Improvements of 10 times would reduce this load by half. The presentation to the panel demonstrated a strong grasp of the landscape and technology options, ranging from custom integrated circuits (ICs), field-programmable gate arrays (FPGAs), and graphics processing units (GPUs) to simply updating designs to more current Si design nodes and III-V technology. Radio range requirements, combined with the digital processing demands encountered in a contested-spectrum environment, present unique challenges for soldier radios relative to commercial telecommunications. In addition to targeting nonconventional mixer-first architectures, this effort is harnessing state-of-the-art 3D architectures available through DARPA's diverse accessible heterogeneous integration (DAHI) project to optimally utilize III-V's for RF power and front-end, with Si for custom FPGAs and dynamic random-access memory for digital and baseband to substantially reduce power.

The project is targeting compelling Army needs and is leveraging current technology advances. The ARL team is working effectively in an industry and academic collaboration that is proving fruitful. Beyond just updating the older soldier technology to current low-power Si, modeling suggests that receiver power can be reduced to as little as 29 to 54 percent of any previously reported designs.

Porous Silicon for On-Chip Energetic Materials

This is an innovative, well-conceived project directed at a clear and unique Army need: Equipment left behind, or provided to a former ally that is now an adversary, is being used by the highly distributed terrorist groups that have appeared in the past decade. Being able to remotely disable any weapon, vehicle, land mine, etc. is a crucial requirement for new equipment. ARL appears to be breaking new ground in this area and needs to ensure that the other DOD laboratories are aware of this basic technology.

Creating a Soft and Stretchable Power System

This project addresses the trade-off between performance and stretchability in power conversion. Inductors have been chosen for study because they are components in wireless power systems, and most features that make an inductor good also make it rigid (thick metal cross sections, parallel traces, magnetic cores). This research involves creating a stretchable magnetic core inductor. The project fits well with ARL campaign plans, including plans to integrate multimodal harvesters with small-scale power management units for low- or no-power sensors and wearable electronics as well as plans to exploit new materials and transduction phenomena for new power conversion topologies. The project appears to be well motivated by future Army needs in wearable electronics.

ARL staff believes that they are at the forefront in area of stretchable power; whereas others are more focused on sensing, displays, etc. ARL noted its strengths in magnetic materials and modeling and design, which is supported by its excellent publication record. The focus is on stretchable magnetic materials (liquid metal inductors and magnetic particles in elastomers) for power components and on modeling and design; this appears to be well differentiated from research at other institutions. Devices have been fabricated and tested, and the publication record for the project is extensive. There is an excellent vision for a range of future applications.

Wide Bandgap Materials and Devices

This project is a new (1 year), fundamental program motivated by the need for more efficient high-power RF sources than are currently available in SiC devices. The work is at a materials and properties level, including performance diminishing factors, such as defects. The researchers on this effort have a good range of materials capabilities and characterization tools available to pursue the development of AlGaN materials and RF devices and good connections to startup companies for collaboration and potential manufacturing, if successful several years down the road. The project is well supported in the materials and devices area but might be aided by adding a staff member with a strong background in the areas of high-power inverters, drives, and hybrid vehicles.

Improved Voltage Control and Stability of SiC Power MOSFETs

This project is a noteworthy example of ARL work maturing into materials processing and devices that can strongly impact Army needs. The effort has discovered a defect source of instability that was completely unknown to the larger technical community and has led to new testing standards for SiC metal-oxide semiconductor field-effect transistors (MOSFETs). Working collaboratively with academic and industry partners, the team has been able to model the physics of voltage instabilities in gate dielectrics for SiC power MOSFETs and promote design and process modifications to address the problem in a commercial implementation. ARL needs to be sure to include examples such as this in future visits because it gives the panel concrete, unambiguous examples of impact. It is also a remarkable illustration of the effectiveness of geographically and institutionally distributed teams in solving design and process challenges in manufacture.

SiC Avalanche Diode

This project addresses an Army need for higher performance protection of electronics (e.g., 600 V vehicle buss) and is leveraging industrial collaboration with Cree, Inc.

High-voltage avalanche breakdown diodes (ABDs) in SiC were shown experimentally to provide improved inductive surge suppression performance relative to commercially available silicon transient voltage-suppression devices. Follow-on work is addressing packaging and higher energy dissipation designs. Beyond publications, the research conducted in this project has successfully established a baseline proof-of-concept to support a commercialization decision by Cree, Inc., to provide SiC ABDs for application in Army kV-level surge environments.

Validated Numerical Model of 4H-SiC Insulated Gate Bipolar Transistor

This project is motivated by high-voltage Army applications (20 to 27 kV) such as fusing, tamper response, and thermal batteries. Because of the higher voltages, special challenges and differentiation include corona effects not seen at lower commercial voltages. Good modeling work is coupled to a foundry that supplies functional chips for high-power switching and systems incorporation. The models have been validated with experiments and predict, for example, temperature profiles in the materials.

Hybrid 3-D Additive Electronics

This is an exciting new capability with wide-ranging applications, including prototyping of new devices and structures and the ability to produce replacement parts in the field. Specifically, this project is investigating the rapid automated manufacture of 3D-additive-printed structures with integrated sensor and other electronic packaging. It is supported by a unique high-precision 3D-additive-process machine that performs extrusion printing, film printing, sintering, and pick and place robotics. The work is well motivated by the Army's need for in-field fabrication and repair, cost reduction (e.g., 3D printed components compared to high-cost machined components), and sensors and other structures on textiles (e.g., for human monitoring). The unique ARL-developed system was made possible by great deal of sound, basic science.

Linear, Efficient 216 GHz InP Transmitter

This project is directed toward a specific Army need based on InP technology—a novel mixer-less transmitter design in high-speed InP circuits that enables higher precision in amplitude in phase to achieve complex constellations despite underlying strong nonlinearities of components. The research team achieved a record 100 mW transmitter power with a spectral efficiency 16-QAM (or quadrature amplitude modulation) format at 216 GHz. This work demonstrates impressive sophistication in RF design at ARL that leverages both strong corporate collaborations for component support and prior ARL work to achieve outstanding results captured in patent applications, publications, and conferences. The devices and amplifiers are all made by outside foundries with good industrial collaboration.

Enhanced RF Device Performance Using Metaferrites

This is an outstanding project motivated by Army communication needs (high-performance, low-profile antennas for vehicles and aircraft; tailored bands). This project provides another powerful illustration of transitioning basic research into novel but practical solutions to address Army needs. Using a novel metamaterials approach in a highly sophisticated layered medium, the international team was able to realize reductions in the required stand-off from a vehicle surface by 85 percent relative to commercial solutions, while also improving the signal-to-noise ratio. Development proceeded from recognition of opportunity of novel materials, to initial evaluation, to refinement, to prototype antenna for vehicle (2″ height versus 14″ height, higher bandwidth, lower losses, etc.). The effort is integrated with other ARL capabilities in RF system design and is now transitioning to industry through the MANTECH program for use by U.S. Special Operations Command to reduce cost. However, there will still be a significant role for ARL to ensure reliable knowledge transfer of the very sophisticated, many-layer structure that needs to be uniform over a quite large area. This project is an example of a successful transition of 6.1/6.2 mission funding.

PiezoMEMS

This project is well motivated by Army communication needs (high performance mechanical filtering in radios including reduced losses and noise) and potential applications for gyroscopes.

The project leverages ARL's leadership in PiezoMEMS research and is well integrated with other ARL capabilities, including fabrication and RF system design and coupled into commercial designs for Army use, such as the common sensor radio. The team is commended for outstanding collaborations across the academic community, engagement in a strong portfolio of industry CRADAs, and commercial licensing of technology. This is also one of ARL's most prolific efforts with 55 publications and presentations in the past 2 years alone. The underlying materials capability can now support sophisticated filter architectures, putting continued pressure on the development of highly accurate modeling.

Opportunities and Challenges

Vertical 2D-3D Semiconductor Heterostructures

There is a significant effort in modeling in this program—which is absolutely critical, although the impact is less clear—because the efficacy of efforts to model vertical transport at van der Waals interfaces using the various available codes, especially in the context of doped materials, has not yet been well developed. The focus is entirely on vertical transport, but lateral transport and novel planar devices might also benefit from this effort.

This innovative program raises many good questions and opportunities for scientific discovery, and also includes valuable collaborative elements with the National Institute of Standards and Technology, Pennsylvania State University, and the University of Arizona that may help to illustrate the impact of ARL's efforts to achieve a more open, collaborative national engagement. ARL needs to give this high-risk program ample time to build a foundational understanding of the technological potential of 2D/3D solutions to address challenging Army needs.

Energy Efficient Soldier Radios

This project spawned extensive discussion of the impact of digital security implementations that have difficulty keeping up with commercial advances, owing to long defense procurement cycles, and the potential that adversaries may achieve an advantage in some circumstances using commercial off-the-shelf (COTS) telecommunications solutions with lower size, weight, power, and cost. With commercial smartphones having a 2-year design cycle, in about a decade, a soldier could be stuck with a 3-generations-old phone compared to the weight, power, and performance capabilities of the newest-generation phones. The Army is uniquely sensitive to this problem, both from a cost and weight perspective, and ARL needs to work closely with the security agencies to quantify these risks and find ways to dramatically accelerate deployments using current commercial design nodes plus FPGAs that might be added to customize and provide security codes that could be changed on a very frequent basis and provide adequate security to commercial designs.

Porous Silicon for On-Chip Energetic Materials

This technology hinges on the ability to integrate fuses, a small destructive charge, with silicon ICs. One application for these ICs is already in transition, a thermal battery supported by a MANTECH program. The researchers have done an excellent job of understanding the system and mechanisms from an analytical and an experimental perspective, which is especially true for the important roles of hydrogen and unreacted amorphous silicon. The investigators need to look for a more environmentally friendly oxidizer and establish its stability with age as well as continue both experimental and analytic studies of the mechanisms and morphology of void formation and oxidizer penetration in the manufacturing process.

Hybrid 3D Additive Electronics

The ability to do 3D microscopy using computerized tomography (CT) scanning is intriguing, and the researchers need to reach out within ARL to see where else it might be useful. One application that comes to mind is the analysis of potential counterfeit electronic parts, which is an expanding problem as IC manufacture has moved mostly to China and obsolescent parts are difficult to find.

The work now is more focused on technological testing and developmental trial on application to Army field-deployed systems. Immediate challenges include integrating G codes for 3D printing assembly robotics that are generated by different software packages; vendors are expected to eventually provide integrated software. The presenter to the panel was a relatively new postdoctoral researcher; hence, the project is only in its very early stage. While it was not clear what the future scientific dimensions of the project will be or what was the anticipated research timeline, it was clear that the project will be relevant to several research thrusts in areas such as alternative energy, novel sensing technologies, lightweighting, and lower-cost manufacture and supply.

Diamond: Novel Material for RF Electronics

This is an excellent research topic because there are major unanswered challenges. This effort addresses an Army need and is supported by good industrial collaboration. Success will definitely be well into the future.

Pristine Layered Materials-Complex Crystalline Structures

This is a long-range effort investigating van der Waals–bonded crystalline layers separated by self-assembled monolayers, with the goal of improving transport and possible foundation for multilevel logic through tunneling between levels created in the ultrathin, multilayer structure. This is a high-risk project with uncertain payoff and demands that evaluation criteria be established by which to assess the continuation of the project.

Although motivated as a device performance investigation, this effort is in fact basic materials processing research focused on deposition and microscopic evaluation of novel, complex, 2D-layered structures. At this early stage, this is understandable, but the team needs to nevertheless select some prototype device architectures that will help to drive the materials configurations under study toward the most potentially useful targets. Overall, this project is at too early a stage and too undefined to assess potential for impact.

OVERALL QUALITY OF THE WORK

Overall, the researchers and the management in ARL's Materials Research Campaign are of high caliber. ARL has utilized short-term (5 years) hiring options effectively to increase the number of early-career staff and postdoctoral researchers who are bringing new ideas, technologies, and enthusiasm to the organization. The level of dedication and enthusiasm was notable and contributed to the creative environment important to scientific and engineering productivity. ARL is to be complimented for following through to establish distinguished postdoctoral fellowships, as recommended by the ARLTAB 2 years ago and reiterated last year. This will undoubtedly attract more top early-career talent.

Most of the projects presented are excellent and have a pervasive potential impact—both short and intermediate term—as well as providing capabilities that will enable the swift response to unanticipated future needs. The scientific soundness and the use of fundamental sciences are outstanding. The project portfolio fits well with both global thrusts and the national agenda, with research projects falling at the intersections of biotechnology, nanotechnology, advanced materials, energy, and the environment.

ARL is making progress in its quest to become a premier research institution in the area of materials science. Several postdoctoral researchers have joined ARL as full-time employees after completing their fellowships, an indication that laboratory management is providing an attractive environment for early-career researchers. It is commendable that the ARL materials science talent pool has a good mix that ranges from experienced, savvy scientists and engineers to bright early-career professionals. There appears to be good diversity with respect to gender and ethnicity.

Collaborative efforts have been demonstrated both across ARL and with external entities. All the projects reviewed are engaged in collaborative efforts to various degrees; this is commendable. The next level of excellence can be achieved by improving the efficiency of this collaboration to deliver better focus, quality, and selection of projects. Internal collaboration across the divisions and directorates is as beneficial as extramural collaboration. ARL's open campus initiative can enhance collaborations.

Advances in biomaterials are essential for applying biology to detection and sensing. The fledgling field of bioinspired and biomimetic materials will be an important source of inspiration and insight for the future materials scientist. This relatively immature ARL thrust is growing rapidly and shows tremendous potential. Because biology is a growth area, ARL has an opportunity to identify and recruit a critical mass of microbiologists and polymer/organic chemists and needs to be looking well into the future to create an integrated community of researchers.

Developing and improving energy storage devices and batteries will be essential if the future warfighter is to gain an advantage from the increasing availability of relevant technology. The same advances will also find applications across a wide nonmilitary spectrum. ARL's research in this crucial arena is broad, covering different devices, fuels, and applications across a wide range of time and size scales. ARL needs to move aggressively to capitalize on internal and external advances in the energy and power arena. For example, the world-leading results on enhancement in QWIP efficiencies need to be translated into capability demonstrators for manufacturers and customers. ARL needs to work more aggressively to leverage external advances in silicon photonics, especially with regard to heterogeneous materials.

Engineered photonic materials are necessary for sensors, energy generation, and improvements to device performance—all essential to the future warfighter. The portfolio of the engineered photonics materials group shows a good balance of high-risk, longer-term work with nearer-term, customer-driven solutions or incremental but critical technology refinement. ARL needs to continue on its course to broaden modeling in support of a larger number of problems and applications. As a prototype for this expansion, ARL needs to look to its short-wavelength infrared device modeling and optimization. The software tool set coming from this research is essential for designers, and it may also provide critically sensitive parameters for potential use in process control for commercial partners and suppliers of imaging solutions to the Army, which necessitates engaging with the manufacturers.

Other methods of high-strain-rate deformation and fracture and shock wave physics experiments that are beyond just ballistic testing need to be explored either through collaborations or development of in-house capabilities. Such experiments are essential in helping to probe material-level mechanisms that can be used for validation of models. Some shock physics experimental capabilities already exist at ARL. It would be valuable to integrate these capabilities as part of the high-strain-rate efforts, while also leveraging the dynamic sector facilities at the Advanced Photon Source. ARL is leading efforts in miniaturized Hopkinson bar experimentation to obtain high-strain-rate material properties. Strategically, technology transfer of this capability could be pursued at various institutions and laboratories in the technical community to help foster novel ideas, and ultimately make this a viable and significant technology nationwide.

With regards to multiscale computations, it would be necessary to integrate physical models in conjunction with the framework being developed, since it may not be a trivial task. There may be opportunities for collaborations in multiscale computations with other laboratories as well as in other fields. Just like the open campus concept is enhancing collaborations and bringing new ideas, embedding ARL researchers at other university and national laboratory campuses, particularly in the area of high-strain-rate deformation, will be equally beneficial. The work on polymers seems to integrate aspects of modeling and simulations, with characterization and testing, and synthesis and fabrication with the unique facilities available. Similar integration was not obvious for the work on metals and ceramics. The research on glass fabrication can be further enhanced by understanding the role of impurities (boron, sodium, etc.) at a molecular level. This is crucial and would promote glass fabrication research using computational material design concepts to obtain an optimum material with the desired (dielectric, optical, and strength) properties for applications other than just for ballistics.

CONCLUSIONS AND RECOMMENDATION

ARL's Materials Research Campaign has two mission elements: (1) to respond to existing and anticipated threats and capitalize on recognized opportunities to protect and enable the modern warfighter and (2) to develop the necessary knowledge base, tools, and capabilities that will allow ARL to respond rapidly to unanticipated threats and opportunities—those of 2035.

Every program reviewed can be described in a space of these two mission elements. For example, the scale-bridging and multiscale modeling of the materials program is intended to develop tools and capabilities that will enable materials design at some point in the future—the first mission element. On the other hand, the energy-efficient soldier radios program is directed toward a specific goal—the second mission element. While one mission element or the other dominates these two efforts, most programs have some component of each. For instance, grain boundary modeling and simulation for lightweight protective materials is a fundamental study of B_4C that addressed the important issues at the root of its poor ballistic performance but at the same time is developing tools that may be usefully applied to the study of other ceramic materials.

The ARLTAB was confronted with the task of assessing programs monolithically, which in general involved determining how successful or how likely the program is to meet it materials development goals. One can imagine that the goals are being met but that opportunities are missed to enhance or broaden the tools or knowledge base, or that the tools available are simply not found to be applicable to a specific problem, although they might have potential elsewhere. In essence, ARL is constructing a number of subsystems that may be integrated and employed to respond rapidly to the unanticipated, but it is for the most part evaluating the performance of the these subsystems within the context of a full system.

It would be useful to consider the product of the Materials Research Campaign at the directorate level as new materials subsystems are created and an understanding of their utility is gained, while the product at the campaign level becomes the expertise to assemble these into an efficient materials development system as well as the ability to determine what new subsystems are required.

Recommendation. ARL should create a structure that places all of its Materials Research Campaign programs within the context of a materials development system. On any project, a researcher should be able to clearly explain how their research fits within this system. ARL should develop criteria that allows for the assessment of individual subsystems of the program (to determining if a subsystem allows ARL to respond to unanticipated threats and opportunities in a rapid and efficient manner) as well as the overall performance of the system.

3

Sciences for Lethality and Protection

The Panel on Ballistics Science and Engineering at the Army Research Laboratory (ARL) conducted its review of ARL's programs on battlefield injury mechanisms, directed energy, and armor and adaptive protection on June 23-25, 2015; and its review of ARL's programs on kinetic lethality, including disruptive energetics and propulsion technologies, effects on target—ballistics and blast, and flight, guidance, navigation, and control on June 22-24, 2016, at Aberdeen, Maryland.

ARL's research into lethality and protection sciences during 2015 and 2016 ranges from basic research that improves our fundamental understanding of the scientific phenomena and technology generation that supports battlefield injury mechanisms in human response to threats and human protective equipment; disruptive energetics; directed energy programs; flight, control, and guidance of munitions; and ballistics, blast, and target interaction programs that address weapon-target interactions and armor and adaptive protection developments to benefit the warfighter. ARL's breath of lethality and protection sciences mission scope work is performed within the Weapons and Materials Research Directorate (WMRD), the Survivability and Lethality Analysis Directorate (SLAD), the Human Research and Engineering Directorate (HRED), and the Sensors and Electron Devices Directorate (SEDD). These directorates work collaboratively to execute their mission of leading the Army's research and technology program and analysis efforts to enhance the protection and lethality of the individual warfighter and advanced weapon systems.

BATTLEFIELD INJURY MECHANISMS

Understanding the mechanism of ballistic injury is essential to the mission of ARL, specifically for protecting the warfighter against traumatic brain injury and extremity fracture injuries. All of the presentations related to injury mechanisms supported ARL's recognition of the importance of this issue. The biggest challenge is bridging the science/engineering gap between the materials science—intensity of soldier protective devices and the biomedical aspects of injury mechanisms, or, more precisely, quantifying the level of mechanical insult leading to significant injury. The program, as presented, is a start to bridging this gap. However, increased commitment of resources will be required for it to become state of the art, where it will have to be if it is to enable the protective devices relevant to the threats of the next 25 years. The program presented is a good starting point, but it needs to aspire to create state-of-the-art models of medical injury. This will require improved coordination with the technical leadership of the field. Understanding the mechanisms of injury to the degree needed to give effective protection is key to improved protective designs. Meeting this challenge is essential to the mission of ARL, and the areas of traumatic brain injury (TBI) and musculoskeletal injury will continue to be the main areas of concern, along with an increasingly sophisticated understanding of how mechanical insult impacts neural function. All models need experimental validation, and all experimental programs would benefit from increased use of statistical data evaluation and statistical experimental design.

Computational mechanics work on battlefield injury mechanisms and human response to threats and on protective equipment, including the mechanics of fibers and fiber composites, are being combined with experimental efforts to characterize, validate, and verify the computational results. This combination of efforts is laudable.

The program in human responses to threats is performed mostly by junior staff, who are pursuing research objectives focused on short- to medium-term objectives. While the staff are capable, the research is generally not state of the art. Studies were described that focused on the assessment of neuronal response injury using a blast tube injury model with cells grown in monolayer on a flexible membrane. The flexible membrane is also subjected to defined strains in order to model the induction of cell injury. Primary end points include cell viability, calcium signaling, and cell morphology. Long-term goals are to develop a mechanistic understanding of neuronal responses in 2D and 3D culture systems in response to well-defined strain fields. Extension of these studies to assess damage to brain tissue, whole organs, and/or tissue-engineered models of the regional damage would be worthwhile. The results of clinical functional magnetic resonance imaging (fMRI) studies of TBI might help to target specific regions of the brain for in-depth analysis. Overall, given the current state of neurosciences and the advances in optogenetics and other techniques, the biological studies described were relatively rudimentary. It is unclear whether on-site senior investigators with expertise in neuroscience participated in this program of study. Collaboration seems to be taking place with one postdoctoral fellow's former senior Ph.D. advisor, but further outreach is needed, including with neuroscience investigators in academia, to augment the military's broader TBI research portfolio.

A second set of studies focused on the assessment of neuronal responses to injury using a micro-explosion model involving cells grown in monolayer submerged in an aquarium-based environment. Primary end points include cell viability, calcium signaling, and cell morphology. Future plans are to study neuronal responses in 2D and 3D culture systems placed within a gel-contained model of a human skull. The quality of the biological studies is rudimentary. The investigations do not appear to include the use of a model system in which the stress fields imposed on the cells have been fully characterized. A gel-containing human skull is an interesting model system, but it will require careful correlation of the estimated in vivo force microenvironment with the in vitro system created in their model system.

A third set of investigations focused on assessing the impact of anthropomorphic variability on the mechanisms of human injury, identifying sites of maximum vulnerability, and determining options for designing improved protective garments and equipment. Clinical computerized tomography (CT) data sets are acquired of soldiers killed in action (KIAs) in order to refine existing computational models of human injury and protection. Collaborative efforts have been pursued with the University of Maryland Shock Trauma Center and other programs. The quality of the scientific studies is high and utilizes an appropriate mix of theory, computation, and experimentation applying state-of-the-art laboratory equipment and numerical models. Extension of these studies to CT and MRI data sets of personnel who are injured in the field but not KIA would be worthwhile. The work designed to predict lumbar burst strength is a start but does not represent the thinking of current investigators in the field. It is necessary that the researchers develop communication channels with research leaders to increase the sophistication and applicability of the approach. While the pig skull work was a reasonable approach, it is unclear if the pig skull is relevant to the critical human skull issues that need to be understood.

The computational efforts in the human biomechanics area are somewhat behind the state of the art in computational mechanics of soft tissue. Specific details include the use of linear tetrahedral elements instead of hexahedral elements for soft tissue response, and there was a lack of viscoelastic properties for material response at high strain rates and high pressures. Further, there was only limited inclusion of the effects of statistical variance into necessary parameters of the computational problem to assess these effects. Such information is essential when computationally modeling humans and humanlike responses. The computational team needs to increase the sophistication of the models appropriate for the problem and needs to interact more extensively with subject-matter experts in human and soft tissue material response in the Army research community and the wider research community.

Overall, the biological programs appear connected programmatically, but they seem isolated from a scientific perspective. There appears to be little synergy or communication between the individual researchers. This lack of synergy is particularly significant to the junior staff, who could benefit greatly from strong mentoring by the appropriate technical communities. They need to become familiar with the current state of the art in their research areas and move quickly to achieve that state. They could also benefit from increased management support to help them learn how to overcome the administrative barriers associated with purchasing supplies and equipment.

To bring the current program to state of the art will require increased coordination of ARL technical personnel with the relevant biomedical communities and the hiring of scientists experienced in the computational modeling and experimental exploration of the effects of mechanical trauma on people, especially in the cases of injuries to the brain and extremities. Current personnel would benefit from experienced mentorship and connections to the field as practiced in university and other government laboratories. The equipment was reported to be consistent with beginning stages of the work and commensurate with the early-career status of the researchers and the brief time (1-2 years) that the program has been in operation. There is little evidence in these projects of the longer-range vision of ARL. The work presented continues to concern itself with current or near-term Army needs.

The research to better characterize the properties of materials relevant to protective systems is sophisticated and mature and is providing the data needed to understand the mechanical performance of protective devices. The project on the ballistic response of knitted materials is a small, well-executed modeling effort that is very relevant and important to ARL needs. While the work is not particularly novel, the results are unique and will be useful for the future design of protective equipment. The study of fiber mechanical properties under very high strain rates is impressive and is likely to provide data needed to better model soft and hard armor design and performance. Nonetheless, the scope of both the experimental and computational programs need to be broadened.

ARL reported a new internal program to study the chemistry and processing of the next generation of protective fibers. This program is supported by newly installed facilities, and it will focus on the modification of existing polymers with additives designed to increase overall performance (nanocomposites) and gel spinning of polyethylene. These represent a reasonable start, but there are other areas of both chemistry (next-generation Kevlar, self-healing materials) and processing (nanofiber production, melt spinning precursors) that need to be assessed as potentially attractive research areas for ARL. The understanding and improvement of polymeric components in protective systems is core to the ARL mission. As with the new programs in biology, mentorship and interaction with area leaders is necessary to ensure that the program is state of the art and aimed at producing materials to satisfy current and future (2040) Army needs.

Overall, ARL is to be commended for initiating programs that link the biology of injury to the materials and constructs designed to protect the warfighter. It is difficult to move into new areas and quickly develop state-of-the-art programs—and to assume leadership. The programs reviewed generally demonstrated technical skill in the chosen areas but often were not state of the art, and they seemed out of touch with the relevant scientific communities. There is an obvious challenge in moving quickly from beginner to leader, but this also provides great opportunity to assess the relevant science and engineering and devise programs to leapfrog to the next level of understanding. The program in battlefield mechanisms, human response, and human protective equipment is conducted by a strong cadre of scientists, and a credible program is under way.

Summary of Accomplishments

Battlefield injuries are an important area of research for ARL, because a better understanding of the mechanisms of injury is vital to improving protective measures. This is especially true for protection of the head, where there is considerable uncertainty about allowable levels of shock, which greatly affects protective options. The research projects presented were appropriate to the problem, and the staff is competent. The projects are short to medium term, which is reasonable for the early stages of a new program. As would be expected in a new research area, there are challenges to be overcome.

Challenges and Opportunities

Current projects are not the state of the art. Work at the cutting edge is difficult to maintain in a small program that does not have the option afforded to larger programs of pursuing multiple approaches simultaneously. Nevertheless, a greater effort could be made to assess the current research in the field and move closer to that cutting edge.

The program seems isolated both within ARL and from the larger outside scientific community. The burdens of being a small, new program in a new discipline within a large organization are many. There are fewer opportunities for constructive discussions and feedback, less chance for synergistic collaboration, and poorer awareness of current developments relevant to their own work. There are administrative burdens associated with procuring materials and supplies that are unfamiliar to the procurement branches of the laboratory. The cumulative effect of fighting through these issues will take a toll on the researchers' time and is a distraction from the pressing needs of maintaining a competitive research program. Management could consider assigning a single administrative contact person, who would become familiar with the unusual needs of the program and, perhaps, act as an advocate for the program within the administrative channels. A long-term vision needs to be developed. The beginnings of this thinking were presented, but they are not yet sufficiently developed to be useful. Such a long-term vision could express a philosophy that helps guide resource allocation and program direction.

DIRECTED ENERGY

The ARL S&T campaign plans 2015-2035 and technical strategy documents[1,2] categorize directed energy (DE) as a focused area under the much broader category of electronic warfare (EW), in accordance with the Army's definitions. The ARL posture designations for both radio frequency (RF)-DE and laser-DE are *collaborate* rather than *lead*. The subsuming of DE under EW and a collaborate-only posture indicate that ARL has downgraded the priority of DE within its technology portfolio from its previous robust effort. The consequence of this status change was evident in the current programs presented: They appear to be a small collection of seemingly unrelated projects. In addition, the current program, with the exception of the project in solid-state laser sources for tactical applications, seems to be concluding soon. Noticeably absent from almost all presentations was any thought of how the operational needs that the current systems were designed to meet would be satisfied in the 2035 time frame highlighted by the ARL director.

In view of the currently fragmented DE program, ARL needs to take a strategic look at the DE area to determine its ongoing priority and refocus ARL's effort with a view to the 2035 time frame. This strategic review needs to include consideration of future capabilities that the Army will need that DE might fill, and what DE capabilities might be fielded by our adversaries for which the Army will need countermeasures. A focused ARL DE program would benefit from a systems-level study addressing future Army missions in which DE could play a role and in which DE effectiveness and alternatives to DE are traded off. In this study, ARL could expand and diversify the laser program to seek avenues for integrating the technology with platforms of importance to the Army. Additional missions for DE could include illuminators; multispectral sensing, identification, tracking, targeting, and damage assessment; electronic protection/ countermeasures for enhanced Army platform survivability against optical and IR guided weapons; and nonlethal weapons. Such a broadly based study is the necessary first step in planning a robust and relevant DE program to address the Army's future requirements. ARL has a significant capability in solid-state laser development—an obvious focus area for the future. In most cases the six projects reviewed met or exceeded the evaluation criteria, which included the following: Does the technology maturation employ appropriate laboratory equipment and/or numerical models? Is the research team properly qualified? Do the facilities and laboratory equipment seem to be state of the art?[3] Are the projects crafted to employ the appropriate mix of theory, computation, and experimentation? Specific concerns about individual projects related to these criteria are included in the following evaluations.

A highlight of the overall program in DE is the project on adaptive and scalable high-power, phase-locked fiber laser arrays. This work is a notable achievement, is recognized as such by the technical community, and appears to be ready for the next step, transition to the field.

The Department of Defense (DOD) recently articulated an electromagnetic (EM) maneuver warfare initiative. While ARL researchers did not reference this initiative, if all the services were to develop joint and independent programs as part of this effort, that could give ARL an opportunity to reexamine its role and strategic opportunities in EM maneuver warfare.

[1] U.S. Army Research Laboratory, *Army Research Laboratory S&T Campaign Plans 2015-2035*, Adelphi, Md., September 2014.

[2] U.S. Army Research Laboratory, *Army Research Laboratory Technical Strategy 2015-2035*, Adelphi, Md., April 2014.

[3] Note that the panel did not visit any laboratories during this year's review, so the assessment of the state of the art of the equipment is based solely on the presentations and briefings.

Research Projects

RF-Enabled Detection Location and IED Neutralization Evaluation

The scientific quality of RF-Enabled Detection Location and IED Neutralization Evaluation (REDLINE) research is comparable to that at leading federal, university, and industrial laboratories, both nationally and internationally. This is a first-class effort with full understanding of, and direct access to, operational needs and with a clear systems approach to reducing technical risks and delivering a successful experimental prototype.

The research program reflects a broad understanding of the underlying science and of research conducted elsewhere. The experimental confirmation of a complex propagation, detection, identification, and predetonation process is impressive.

This project is ready to begin the next step, deployment in the field. There is still an applied research effort needed to investigate detection, identification, and predetonation of increasingly advanced, emerging threats. The poster presenters mentioned the potential for mounting the capability on unmanned aerial vehicles. This seems to be a good idea, especially if ARL seeks to investigate the evolving improvised explosive device (IED) threat beyond the near term.

Hostile Fire Detection

In general, the scientific quality of the research is as good as that achieved at leading federal, university, and industrial laboratories, both nationally and internationally. The investigators used standard codes and modeling techniques. Although not strikingly novel, the work was credible and demonstrated useful integration of known techniques. The investigators also appeared to have access to intelligence about specific threats that may not be widely known.

The research program reflects a broad understanding of the underlying science and research conducted elsewhere. The researchers have addressed the major issues associated with detection and geolocation of threats such as rocket-propelled grenades and small arms. There was an appropriate level of modeling and predictive work to address near-term deployment but not longer-term strategic innovation. The prototype work that has been exercised in limited deployment responds to a near-term problem. Advanced (2035 horizon) modeling, diagnostic, sensor development, and test capabilities were not brought up.

Operational data from full field deployments would drive next-generation innovation and improvement in identification and geolocation signature analysis for targeting support. This would produce results that could ultimately be transitioned to the field in a continuous upgrade process.

Adaptive Techniques for Advanced Radar Tracking and Optimization

The scientific quality of the research is basically sound in the context of unclassified university research, but it is not up to the standard of leading federal, university, and industrial laboratories working in this area. There appeared to be little or no awareness of existing, similar work in advanced radar development other than some unclassified university research. Reaching out to a major radar program, perhaps one of the Army's programs, might have revealed similar, prior work and identified what is and is not already in existence.

As for appropriate laboratory equipment and numerical models, there appear to be adequate computing resources but no association with radar R&D facilities or laboratories to ensure a practical base of experiment and experience. It is also not clear whether the signal interference modeling is relevant to existing radar clutter, interference, or jamming environments. Such interference can depend on the design characteristics of the radar under consideration, so general approaches may not be directly relevant.

There could be projects that, with improved direction, access, and resources, produce results that can be transitioned ultimately to the field. Possible collaboration with the Navy's extensive efforts in sonar tracking and optimization may be fruitful. The freshly conceived algorithms and use of greater computing power might provide useful insights to radar R&D facilities and developers. Some algorithms may be interesting for specific interference waveforms as spectral crowding increases.

Solid-State Lasers

The scientific quality of the research is comparable to that achieved by leading federal, university, and industrial laboratories. This research is aimed at identifying candidate materials, methodologies, and techniques for scaling solid-state lasers to mission-significant powers within the constraints of space, weight, and power. Although many laboratories are doing similar work, ARL is concentrating its effort in eye-safer spectral regions that are of critical importance for the Army. The research program reflects a broad understanding of the underlying science and of research conducted elsewhere. ARL's work is known and respected by laser scientists at other institutions.

Programs crafted to employ more modeling would provide an enhanced mix of theory, computation, and experimentation. Given the objective of this project, the researchers need the capability for simulating, even crudely, an entire system from wall plug to target. This is the only way an analysis of alternative materials and architectures can be performed. Such an analysis would permit more informed choices for R&D paths to follow.

Adaptive and Scalable High-Power, Phase-Locked Fiber Laser Arrays

This research program is devoted to developing high-power (tens of kilowatts) fiber lasers by coherently combining lower power systems. The researchers have successfully combined seven lasers, each of which can continuously produce as much as 1.5 kW. A novel method has been developed for coherently combining the multiple beams. This is the critical element of any high-power fiber system. Feedback from a diffractive element located at the output aperture provides an optical signal that serves to phase lock the laser array. Another strength of this method is the modest bandwidth requirement for the feedback system (only 15 kHz), which is very attractive from the perspective of developing a reliable weapons system. Coherently combining the individual laser beams occurs at approximately 10 m from the output aperture, which is well into the far field. Beam quality is also actively monitored in the far field so as to optimize the efficacy of the phase-locking process.

The impact of this ARL laser system appears to be significant. In follow-on work, the Defense Advanced Research Projects Agency and the Lincoln Laboratory of the Massachusetts Institute of Technology have scaled this system so as to combine as many as 21 lasers. Although it is not clear at this point whether the ARL system will ultimately be incorporated into a real weapons system, it is evident that the system architecture has influenced other work. Low-power versions of the ARL design are, for example, being developed for civilian use. Another impressive aspect of this program is that it has resulted in six patents.

The high-power fiber laser system is the result of a decade of work at ARL. This program demonstrates the value to DOD of investing in novel research over a prolonged time. A further accomplishment is the understanding of the physics of intense optical fields propagating in a fiber. One practical outcome of this understanding was the finding that fiber core diameters as large as 20 μm could be used while maintaining beam quality.

Nonlinear Propagation and Target Effects of Ultra-Short-Pulse Lasers

This basic research project examined nonlinear propagation in the atmosphere of an ultra-short-pulse (1 psec) laser beam in a self-generated, ionized channel. The researchers observed that the ionized channel through which the beam propagated was much more stable at a pulse repetition frequency (prf) near 1,000 Hz than at a frequency of 50 Hz. The causal physics was conjectured to be that the channel remained steady at the higher prf owing to the lack of thermal dissipation of energy of the nitrogen and oxygen plasma that formed the channel.

Also reported was that the beam, when incident on solid surfaces, created ripples in the surface of the material over the area covered by the beam. This phenomenon was previously reported by others for metals and semiconductors but was demonstrated for the first time on polymers by the ARL team.

The researcher showed a strong knowledge of the experimental laser techniques and knowledge of previous literature. It was not made clear, though, why the experimenter followed the path he did. The quality of the work appears to be high and the facilities used at ARL were adequate for investigating this phenomenon. It was not clear, however, if computational modeling was performed to substantiate the proposed model. The experimenter did not have a clear idea of where this work was headed and how the Army might benefit from it.

Summary of Accomplishments

The REDLINE team has developed a kill chain concept for the detection, geolocation, identification, and triggering of IEDs. Model predictions and prototype experiments verified the performance of the harmonics-based approach, and the program has advanced to early system prototype testing.

Investigators working in the hostile fire detection area have developed diagnostic, modeling, and prototype hardware capability of detecting and geolocating hostile fire for enhanced soldier survivability. The work addresses three major areas in disrupting the lethality chain: threat signature characterization and identification; analysis of intervening and interfering material; and sensor systems response.

Significant field testing has enabled the development of a large, well-understood archive of unique multispectral data that was used to construct databases for rapid threat identification. ARL's work expands and improves the database of medium wavelength infrared (MWIR) and ultraviolet (UV) threat information. The analytical tools available to model EM propagation through both the atmosphere and various types of obscurants employed fundamental, well-understood concepts. The analytical tool for modeling intervening media and obscurants is a unique capability that was developed with academic collaborations and was empirically validated. The models have been integrated with various types of sensor payloads and packaged into the prototype hardware. The work has produced a patent on optical gunfire rocket and explosive flash detection that has been embedded in the electro-optical (EO)/IR sensor hardware. The investigators have taken the work from innovation to field prototype.

In the program developing adaptive techniques for advanced radar tracking and optimization, the concept involves a radar pre-look at the spectral signal environment prior to each dwell and uses algorithms to select quieter frequency gaps to form appropriate waveforms that minimize received interference while retaining required waveform resolution.

This experimental work on adaptive and scalable high-power phase-locked fiber laser arrays was outstanding; the experimenters clearly understood the issue and why it was being pursued, and they described well the problems that had to be overcome to produce the results of this beam combining experiment. Given the available laser power, the results were impressive and are headed in the right direction for producing a high-quality ($M^2 \sim 1$) combined beam from 6 to 8 fiber lasers that are all phase-controlled using an innovative optical feedback technique. Effective use of laboratory equipment was demonstrated. Whether or not this work can combine a sufficient number of fiber lasers to produce a 100 kW class laser is not clear.

Challenges and Opportunities

Limited test results of the REDLINE team confirm theoretical predictions of range, detection, and identification. However, an ROC curve (probability of detection versus probability of false alarm) based on test results and model predictions is not yet complete. Similar test data are needed for the likelihood of killing an identified target. Such a comprehensive characterization is needed, especially in a cluttered urban environment, as part of the program to verify that the system is operationally viable. This information will be required if the range of the system, say, by utilizing an unmanned aerial vehicle, is to be considered.

Also, the REDLINE team indicated an upcoming transition to 6.3- and 6.4-funded development. However, there is still 6.2-funded R&D to be performed, including characterizing emerging trigger threats and other countermeasures and design modifications to accommodate those evolving threats. Because the IED threat is expected to continue, a critical need exists for a continuing research program to address the projected and potential advances of the threat in the coming decades.

No strategic plan was presented for further development and maturation of the models for hostile fire detection, for advanced sensor capabilities, or for continuing experimental evaluation of future threats. To be effective, contributors and researchers need to become involved with established radar S&T and R&D groups—for example, such groups within the Army—to gain feedback on the viability and value of the approach compared to earlier work. The qualifications of the research team in the area of adaptive techniques may not be up to the research challenge, given the team's lack of access to operational radars and ongoing radar developments. There does not seem to be a core radar group within which this work is performed, so it is unclear why the work is under way in this particular research campaign. This may be a strategic question for ARL relative to the Army technical infrastructure. (The Naval Research Laboratory has had a robust radar research program for years.) The facilities and laboratory equipment may not be state of the art compared to the signal processing laboratories of advanced radar programs. Indeed, there appears to be no radar test site, data collection capabilities, or other laboratories associated with this work. The program is not crafted to employ the appropriate mix of theory, computation, and experimentation nor was there a connection to any existing or new radar and radar R&D facilities.

ARMOR AND ADAPTIVE PROTECTION

ARL has a strong record of achievement in the basic and applied sciences and the engineering of penetration and protection. The ongoing work described in the review showed how ARL is building on this tradition of excellence to provide the knowledge basis for future Army needs. This is a core competency that underlies Army capabilities.

The presentation on penetration, armor, and adaptive protection provided an impressive overview of ongoing research aimed at meeting shorter- and longer-term issues. The shift of focus from the goal of addressing short-term Army needs to the goal of carrying out research that will maintain world leadership in this area for future Army needs was evident.

The depth of knowledge of the staff and the evidence of interaction between staff members were impressive. They were also aware of and knowledgeable about projects other than their own. It is important that ARL ensure a steady supply of new staff into this critical area and that newcomers can benefit from the experience of senior researchers.

There was significant evidence of teamwork and integration among the projects in, for example, adaptive protection. There were examples of linkage of experiments and computational modeling to provide physical insight into problems, to aid in new designs, and to explore new concepts. The combination of modeling and experiments is essential in many cases, but there are circumstances in which it is appropriate to focus on a single mode of inquiry: experiments carried out as discovery science; modeling to develop an understanding of scenarios that are impossible or prohibitively expensive to investigate experimentally; development of new modeling approaches and techniques that promise to enhance predictive capabilities of ballistic phenomena; and development of new experimental methods that promise to provide a better understanding of the physical mechanisms underlying ballistic phenomena. ARL described a ceramic armor concept that was made possible by a previously developed experimental technique aimed at enhancing a basic measurement capability.

The staff apparently have freedom to pursue new ideas that can lead to breakthroughs that might otherwise be found more slowly, if at all. An example was the armor concept, a serendipitous discovery developed nearly to completion before being fully funded.

Developing a predictive capability for damage and fracture in metals, ceramics, and polymers underlies the efficient development of new material systems for protection and for penetration. At present, there is no framework that has penetration capability. However, experimental, theoretical, and computational advances being worked on in other countries are making such a capability seem possible in the not-too-distant future. A systematic approach based on understanding the key physical processes is needed because of the wide range of material systems that are becoming available. There are so many possibilities that a trial error-and-correction approach would be too expensive. It is important that ARL develop a leadership capability in this area. That requires the ability to identify damage and failure mechanisms in material systems of interest, the theoretical expertise to model these failure mechanisms, and the computational ability to simulate armor concepts and designs for the range of conditions encountered in the field. It is unlikely that a detailed quantitative capability will be developed. A more realistic expectation is a predictive capability that ranks the response of proposed armor systems to various threats and provides scaling relations that can be confidently used to transfer laboratory-scale tests to field condition response. Success in this area requires hiring and developing a critical mass of staff and having the needed experimental and computational capabilities.

Modeling

As pointed out above, ARL uses both experiments and modeling to develop new armor concepts and designs. ARL's use of modeling is maturing and is becoming better integrated into armor development and design. The researchers presented evidence that ARL was using numerical simulations to explore armor concepts more expediently than could be done through experiments. There were also examples of modeling being used to provide physical insight into experimentally observed phenomena, and there were examples of concepts and designs being examined that could not be tested experimentally with current capabilities.

Numerical simulation represents a key capability for ARL in the armor and adaptive protection area. ARL staff are customers for and collaborators with developers of advanced computational tools. Much of this activity involves codes developed at Department of Energy (DOE) National Nuclear Security

Administration laboratories (Lawrence Livermore National Laboratory [LLNL] and Sandia National Laboratories [SNL]). These tools include ALE3D (LLNL), ALEGRA (SNL), and CTH (SNL). Some usage of multiphysics Sierra codes (SNL) was also reported. These are probably the appropriate tools for ARL's problem set (impact, high rate, energetic materials, and electromagnetics), because they scale well on parallel platforms and are the most advanced tools available. There was some use of commercial codes (e.g., LS-DYNA) as well, which allows ARL to exploit developments in, for example, crash-worthiness analysis as it relates to the automobile industry.

ARL's relationship with the ALEGRA and CTH development teams at SNL has allowed it to drive the code development to address its own needs. ALEGRA is an arbitrary Lagrangian-Eulerian code with electromagnetics capabilities that is well-suited to a specific subset of ARL's problems. ARL staff are trained in use of the code, and this seems to have improved the sophistication of the analyses conducted. ARL is a significant user of CTH (SNL Eulerian shock physics code) for armor and adaptive penetration applications; in fact it is perhaps the largest DOD user as measured by central processing unit hours. This is ARL's workhorse code for impact problems. ALE3D is utilized for these problems as well. ARL staff members develop constitutive models to describe material behavior for all of these codes, which speaks to the level of sophistication of ARL modeling.

There was some evidence of the use of multiple codes to address different physics in a single problem. Use of the codes in this way will likely increase in the future, although coupling of codes is a challenging endeavor that will make the development of general frameworks for the coupling of codes increasingly useful.

ARL researchers indicated that their overall framework for multiscale modeling is also intended for armor and adaptive protection problems. The multiscale modeling work will likely become increasingly important for modeling complicated material behavior.

There was evidence that the researchers' computational work was limited by the available classified computing capability. ARL indicated that a 100,000 (node or core) machine was available for unclassified work but only a 15,000 (node or core) machine existed for classified work. For 3D magnetohydrodynamic calculations with ALEGRA, thousands of cores are required for several days—a significant portion of the computing power available at ARL. ARL therefore does much work of this type in a 2D axisymmetric configuration. Although this is less computationally expensive and is useful for many problems, it limits ARL's capability to explore oblique impact conditions and other scenarios that are not axisymmetric. Also, as ARL works to develop its parametric studies and its verification, validation, and quantification of margins and uncertainties (V&V/QMU), many more simulations will be required, further straining the available computing power. ARL needs or will soon need more powerful classified computational platforms in order to accomplish its mission. A challenge in justifying more powerful classified machines is that ARL's relatively small classified user community places high demands on the machines at some times and lower demands at others, potentially leaving significant portions of a large computing cluster idle. A potential solution to this is to utilize designs that allow sections of a large computing cluster to swing between unclassified and classified mode. In this manner, the allocation of resources can more effectively address the needs for the two types of computing resources.

Developing predictive models for damage and fracture for armor and adaptive protection applications is an important research direction, and in these circumstances the material response is not likely to be entirely deterministic. Therefore, the scientific and evidentiary value of this research effort will be greatly enhanced by adopting a ubiquitous statistical perspective. Understanding the nature of the assumptions and approximations underlying predicted or anticipated behavior and how these can be updated as data/knowledge is gained will improve ARL's ability to develop technologies to adapt and survive in extreme and hostile environments. Furthermore, statistical scatter in experimental data could be an indication of subscale behavior with implications for modeling, so its impact on predictions needs to be explored through sensitivity studies and uncertainty quantification methods.

Experimental Aspects

ARL's work in armor and adaptive protection is also supported by experimental work. ARL utilizes its in-house capabilities for ballistics testing, which appears to be fairly well developed. Nonetheless, ARL needs to develop a wish list for experimental capabilities as well as a timetable for obtaining them for future needs.

ARL is also utilizing unique national facilities such as the Dynamic Compression Sector at the Advanced Photon Source at Argonne National Laboratory (ANL) and the proton radiography (pRad) capability at the Los Alamos National Laboratory (LANL). Utilizing advanced facilities in this manner will advance ARL's science base and leverage these important national capabilities.

There were also instances in which ARL identified important technical developments and brought them to ARL. For example, it is developing a flash tomography capability and a capability to utilize photon Doppler velocimetry (PDV) in its work. It is important that ARL continue to find important technological developments and bring them to ARL when appropriate. In the case of PDV, ARL would benefit from engagement with the wider PDV community (e.g., the PDV workshop) and, if possible, seek out a short course that would train staff in the use of the PDV. ARL will also need to figure out how to exploit PDV effectively in its work.

The panel encourages continued development of the relationship with the additive manufacturing group at ARL and with experts around the country. Additive manufacturing has the potential to enable new armor concepts but could at the same time lead to new threats from adversaries.

There was significant discussion of the use of energetics to solve armor and adaptive protection problems, but there was little discussion of the science of energetics. The armor and adaptive protection group needs to engage more with the energetics group at ARL as well as with outside experts. For example, there are several concepts that rely on modification of explosive sensitivity that may be beyond current ARL capabilities. Technologies being developed in this area have the potential to enable significant advances in armor capabilities. Furthermore, state-of-the-art tools for modeling energetic materials are being developed elsewhere at ARL that may be applicable to armor and adaptive protection problems. The science of energetics in the context of armor and adaptive protection may be significantly different from that science in the context of warheads, so the ARL group working on armor and adaptive protection may benefit from a workshop on energetic materials for reactive armor. They might also encourage the Army Research Office to establish a Multidisciplinary University Research Initiative in this area.

Summary of Accomplishments

ARL has a strong record of achievement in the basic and applied sciences and the engineering of penetration and protection. Its presentation of experimental and modeling results and progress in penetration, armor, and adaptive protection provided an impressive summary of ongoing research aimed at meeting short- and longer-term mission needs.

There was significant evidence of teamwork and integration among the projects, in, for example, adaptive protection. Examples of the connection between experimentation and computational modeling that gave physical insight into problems were especially noteworthy; such work is likely to aid in developing new designs and exploring new concepts.

The benchmarking of simulations with experiments and the emphasis on bringing advanced technology (particularly in the X-ray region) to bear on diagnostics were impressive.

Challenges and Opportunities

One challenge for those working in applied classified areas of armor R&D is to figure out ways to interact with outside experts. Ways to do this include participating in appropriate forums (classified meetings, interlaboratory workshops, international exchanges); identifying canonical unclassified problems and cases that can serve as conduits for collaborations with universities and other outside experts; and conference participation, which is very important even for those who cannot present because their work is classified. Conference attendance by those working in classified areas helps them remain up-to-date in their fields.

Rigorous procedures for the validation of model-based predictions that are consistent with current state-of-the-art methods use experimental data and the propagation of uncertainty as well as the characterization of associated modeling errors. This requirement is exacerbated by the complex multiscale and multiphysics interactions relevant to many predictive efforts that are under way at ARL in the armor and adaptive protection areas.

As ARL works to develop its use of parametric studies and V&V/QMU, many more classified simulations will be required, further straining the available classified computing power. ARL needs to elucidate a strategic plan for more powerful classified computational platforms in order to accomplish its short-term and current mission needs and to support future mission needs and deliverables. It also needs to continue development of its relationships and projects examining the utilization of additive manufacturing (AM) to address current and future Army needs. There is an opportunity for the ARL additive manufacturing group to interact and collaborate with experts around the country at DOD facilities and federal agencies and in academia and industry. Additive manufacturing has the potential to enable new armor and protection as well as new weapon concepts; AM could also lead to new threats from adversaries, which means new challenges to our warfighters. There is a need as well for procedures to qualify and certify AM materials to meet Army needs. AM has become a realm where new ideas are being developed and where the future Army is being enabled, so that ARL could become involved in AM work, and ARL needs to develop a strategic plan in this area. ARL's modeling programs could embrace the importance of variations, errors, and margins for establishing thresholds and statistics that support the development of predictive capability and design capability.

The presentations on damage and failure modeling demonstrated that modeling of damage evolution, fracture, and failure is a critical prerequisite for developing predictive and design capabilities in penetration mechanics. It is critical that ARL establish a focus in this area as soon as possible.

DISRUPTIVE ENERGETICS AND PROPULSION TECHNOLOGIES

The disruptive energetics and propulsion technologies reviewed in 2016 highlighted research and technology advances in four areas: synthesis activities, propellant simulation, extended solids, and multiscale computational modeling.

Summary of Accomplishments

Synthesis Effort

This is a commendable, relatively new effort to develop a chemical synthesis effort at ARL. Other laboratories have moved away from this important area for the Department of Defense (DOD) and the country as a whole. Synthesis efforts stimulate many other efforts at ARL—new materials challenge and sharpen skills in formulation, characterization, and testing activities. Some new molecules were synthesized, which is the first from ARL for many years. This is a milestone for ARL and the research area. The particular work presented to the panel represents a good research direction (specifically looking at nitroglycerin replacements) that could yield better migration behavior eventually. It is notable that materials are being sent to Aviation and Missile Research Development and Engineering Center (AMRDEC) and to academia, showing good collaboration and interactions. There is a good focus on all possible applications (propellants and explosives), rather than just explosives as a main focus. There is also consideration of what the ultimate formulations are (how molecules might interact with nitroglycerin, for example), the possible toxicity of their materials, and not just making a new molecule to publish. This team wants to make materials that make a difference to the Army.

Propellant Simulation

This area is a traditional strength of ARL that positively impacts and supports Army needs (and other research efforts) well beyond ARL. It needs to be vigorously supported and maintained. Effort is needed to execute succession planning to sustain ARL's strengths in this area as staff retire.

The propellant simulations presented suggest that the observed propellant burning rate slope break is from the interaction between mechanics and reaction. This is an important result and has significant implications. Basic combustion theory supports this. Specifically, the burning rate depends on pressure raised to the overall reaction order divided by two ($r_b \; \alpha \; p^{n/2}$). Therefore, for overall reaction orders of two ($n = 2$) that are typical, the pressure exponent is about 1. Therefore, above the slope break, where exponents are well over 1, it is highly unlikely that the slope break is chemical in origin, as some have suggested previously. This work also suggests future directions for research. For example, the oxidizers or binders could be chosen to mitigate fracture. Also, particle size or morphology (even porous oxidizer perhaps) might affect the slope break phenomena. There is good interaction with experiment and modeling in the embedded wire propellant project.

Extended Solids

This is an ambitious and high-stakes project. It is the most high-risk/high-payoff project of the extended solids program that was presented to the panel. It is commendable that some materials have been scaled to more reasonable quantities (grams), but scaling remains a significant challenge. The issue of stability (could be extremely stable or unstable) is another challenge and is an area where theory and modeling can play a major role. In some projects, theory and experiment were well connected, but some projects had evolved to an exploratory mode with little theory guidance.

The small-scale characterization experiments (laser shock, electrostatic discharge, specific impulse or Isp from strand burns) are in need of a more scientific underpinning to connect with more than 10 g sample properties. A detailed analysis of these experiments is needed. For example, can it be shown that shock speed in the laser shock experiment could be related to detonation speed? Also, can these small-scale characterization experiments be repeated by other groups—a basic requirement of scientific advancement? How well do all these approaches work for standard materials where large-scale experimental data is available?

Multiscale Computational Modeling

A fundamental weakness of macro-scale continuum simulation methods in heterogeneous energetic materials is the loss of detailed material information at scales below the resolution of computational elements or cells. Homogenization is acceptable when average response is desired on the relevant length-time scales, or when validation is carried out with experiments that similarly measure average properties. However, problems like explosive initiation and impact thermal runaway, or specifically explosive failure, requires the prediction of rapid dynamic phenomena operating in highly unstable threshold regimes, where subgrid material structure, such as porosity and multipoint material correlations, dominate the transient response. It is precisely these regimes that need to be characterized in order to understand performance and safety margins, and the uncertainties stemming from process and manufacturing variability. ARL is making good strides in this area and is encouraged to continue to broaden its efforts. A transformative capability that predicts dynamic response on component scales of interest via comprehensive treatment of the scale relevant physics will truly lead to an improved understanding of the response of new and traditional energetic materials.

To this end, novel tools and analyses that reveal the complex dependencies and correlations between nonlinear transient processes in complex microstructures and the underlying stochastic variation of microstructure are needed. This understanding could be used to develop a statistical description of heterogeneous materials that can be incorporated into macro-scale continuum methods that embed mesoscale response as suggested by lower scale (mesoscale) simulations. The correlation of grain scale microstructures and material shock response is the key and essential feature of bridging scales to the continuum level.

Challenges and Opportunities

In the development of improved propellant energetic materials, more interaction with computational modeling efforts is needed, and it appears that that is already starting. Also, this group has plans for several new molecules and seems to have a good plan to achieve the results. This could be a growth area for ARL and is encouraged.

There are plans to do phase-contrast X-ray experiments at ANL in support of the extended-solids projects when a new facility is available. Some of these experiments may be attempted with current capabilities at ANL; ARL's staff is encouraged to explore what experimental aspects of their research could be started now with the current facilities at ANL.

There is good interaction between experiment and modeling with the embedded wire propellant project, although there was not time in the presentation to the panel to show this, and it was not in the experimental poster. Future work in this area might consider reactive wires or printed wires.

In the effort to address the scales greater than atomistic quantum levels, dissipative particle dynamics has been demonstrated as a promising approach for coarse-grain numerical simulations. Nonetheless, it still represents an intermediate scale toward coupling to the continuum level. Although ARL researchers are well aware of this intermediate level, and perhaps the sequential hierarchical approach may be a good method to bridge scales, it just needs to be more complete. A clearer roadmap for a better picture of bridging the scales to the continuum would have given a clearer way of defining where, ultimately, computational modeling for reactive materials is headed.

A plan for validating the multiscale modeling was put forth based on continuum-level testing and diagnostics. This has been the traditional approach in assessing modeling at the continuum level, but it is still removed from validating the bridging of scales. There is a vision forward to address material strength effects based on continuum-level constitutive modeling; however, the link to dissipative particle dynamics (DPD) models still remains questionable. In the effort to address the scales greater than atomistic quantum levels, DPD has been demonstrated to be a promising approach for coarse-grain numerical simulations that implicitly couple to molecular-length scales.

An area where continued attention needs to be focused, and which represents a great opportunity, is investing in the characterization of inter-material interfaces. The study of interfaces provides opportunities for bridging length and time, approaches to how these scale more effectively (possibly with statistical or probabilistic treatments), and seeking connections (validation, insight, etc.) to experiments that reveal subscale behavior are needed.

EFFECTS ON TARGET—BALLISTICS AND BLAST

The teams working on materials synthesis and propellants demonstrated high technical competency and in-depth understanding. The teams are commended for making progress in technically challenging areas related to the development of new energetic materials, propellants, and blast effects on targets. The teams clearly demonstrated their understanding of the necessity for experimental validation and are making significant progress. ARL is encouraged to continue move in this direction. ARL is complimented for putting forward some of its early career researchers in this topic area. They showed strong technical competency and poise.

Development of New Energetic Materials

The efforts at synthesis of energetics are cutting-edge work and are showing results in the newly developed promising chemicals. This is a high-risk/high-payoff effort, so ARL could expect that most candidate materials may not, ultimately, transition to Army applications and systems.

One significant challenge in the development of new materials is evaluating or screening candidate materials when only small quantities are available. There is some promising work to develop approaches for testing and screening candidate materials utilizing state-of-the-art diagnostics such as PDV. ARL needs to complement the experimental efforts with modeling efforts that might suggest alternate geometries (e.g., cylindrical) and perhaps allow additional information (e.g., model parameters) to be obtained from the data.

Propellants

The modeling of solid propellants appears to be state of the art. Currently, the reaction modeling approach uses laminate/mixture theory. Good progress has been made in understanding detrimental formation of cracks and the sudden nonlinear increase in burn rate as a function of pressure (referred to as the exponent break). It was shown that this break is not due to chemistry alone (an increase in gas-phase reactions as a function of pressure); rather, it appears that it is due to mechanical processes—for example, the growth of microcracks that increase surface area. There is a desire to push the exponent break to somewhat higher pressures to permit extended range. This enhanced understanding of what causes the break could lead to possible reformulations for improved performance. Validation of the modeling effort is ongoing, including, for example, experiments with a wire or foil insert. Although it was acknowledged that real propellants are heterogeneous, the modeling effort continues to focus on the laminate/mixture theory approach. The current approach could be maintained since it continues to provide fundamental insight. However, ARL needs to pursue a complementary modeling approach that includes the material heterogeneity, because this will ultimately be required to make further progress on the problem.

Energy Release—Target Effects

ARL staff showed recognition that a better understanding of chemical kinetics is required for advancements in solid rocket propellant technology (e.g., propellant reformulation to increase performance). ARL has a computational effort to calculate decomposition and energy release during deflagration, including the impact of additives. The thermochemical parameters and reaction path parameters are being estimated via quantum chemistry modeling, and the researchers are building a library of chemical kinetics reactions and mechanisms. This is a worthwhile endeavor and needs to be pursued. However, effort to validate the modeling effort through experiments was not described. If there is not a parallel effort to support some sort of validation, one needs to be initiated.

A study to investigate the shock energy absorbed by a target appears to be primarily experimental, although the study did show a complementary numerical simulation. The focus is to develop the diagnostic capability to quantify the details of shock impingement. The experimental procedure has the capability to measure air shock speeds as well as see ground reflections and a secondary shock (presumably the result of a collapse of the initial shock and a spherical collapse and rebound of the initial shock's interaction with the explosive/air interface). However, the computation effort focused on an ideal spherical shock expanding and impinging on the target. The researchers also showed the shock impinging on an inclined plate, and the experimental technique could easily quantify the complex shock interactions. With the wealth of experimental data, there could be more of an effort to simulate the actual experiment to validate the modeling capability. The investigators however are disconnected from the potential users of a validated computer code. In addition, the investigators show a shock impinging on a helmet with a polycarbonate shield but showed a 25 psi shock impinging on the helmet. No soldier could even come close to withstanding such an overpressure. So, a different question is this: Can the experimental technique resolve shock data at values that would represent aggressive but survivable loading of a helmet? To investigate interactions of face plate, helmet, TBI, and other factors, the loading conditions need to be accurate. Thus, the effort is worthwhile, but the focus needs to be adjusted to be relevant.

A relatively new study is focused on developing a diagnostic capability to interrogate and measure the very beginning of a reaction. The phenomena, such as electronic transitions, occur on a time scale of femtoseconds to nanoseconds. A significant amount of time was involved in ordering equipment and setting up the experiment, and the initial results look promising. It would probably be useful to initiate a parallel modeling effort that might help to interpret what is being observed.

It is well known that energy release rate is dependent on the surface area of the burning propellant. A desire is to fabricate an optimal shape propellant using 3D additive manufacturing techniques. Modeling is being used to develop appropriate geometric web designs to tailor the burn rate to achieve the ideal constant breach pressure response. Additionally, the modeling includes thermal effects (coefficient of thermal expansion) since the burn rate is a function of initial propellant temperature. As part of the technical effort, the coefficients of thermal expansion were determined for several propellants and celluloids. Part of the design process is addressing the question of whether temperature-sensitive glues could occlude surface area at higher temperatures. It is not clear at this time if the goals of the technical effort—for example, developing 3D additive manufacturing techniques applied to solid propellants and temperature-sensitive glues to control surface area—will be achieved in the remaining year of the project.

Summary of Accomplishments

The teams working on materials synthesis and propellants demonstrated high technical competency and in-depth understanding. The teams are commended for making progress in technically challenging areas. The teams understand the necessity for experimental validation and are making significant progress, although more work needs to be done. The efforts at synthesis of energetics are clearly cutting-edge work and are showing results in the newly developed, promising chemicals.

In the experimental characterization of small quantities (few grams) of the new energetic materials being developed, the fidelity of the experimental data may permit development of energy release models. This is an exciting direction of research, and promising opportunity, supported by ARL's thrust in this area.

Challenges and Opportunities

Some promising work is ongoing to develop approaches for testing and screening candidate energetic materials. However, it remains a significant challenge to develop useful ways to characterize energetic materials to assess if they are promising when only smaller quantities are available (under 2 grams). Synthesis chemists would benefit from working with materials scientists to generate useful characterization information. The analysis of new materials development for propellants and ballistics applications needs improved collaboration with the Materials Research Campaign which could help improve progress in the development of better propellant materials.

Another important challenge is the scale-up path for new materials. In at least one case (plasma-assisted deposition of carbon oxide materials), the ARL team appeared not to have given much consideration to the scale-up process. Because the scale-up process is likely to be difficult and can keep new materials from being useful to the Army, ARL teams need to include scale-up considerations in their work as early as practical.

ARL needs to take a more systematic approach to characterizing the physical (microstructural) and chemical properties of the new materials that have been synthesized through techniques such as Raman, nuclear magnetic resonance, and TEM. Enhanced characterization represents a potent opportunity to support predictive model development with microstructural data and validate the microstructures and chemistry produced in the synthesis efforts.

In some of the research projects, minimal computational components were visible. In the solid-propellant burn rate enhancement for missile propulsion research, where both experiments and modeling were done, no direct comparison between the two was shown. The ARL researchers need to include both experiments and simulations in their research projects.

FLIGHT, GUIDANCE, NAVIGATION, AND CONTROL

Guidance, navigation, and control (GN&C) are what elevate an unguided projectile to a smart munition or a guided missile and are more focused on the objective of reducing the expected miss distance, increasing the reliability and accuracy of engaging the target, and, finally, increasing lethality. Precision on the battlefield remains a potent game changer, and the development of better techniques is fundamental to the Army's future success. The function of navigation and guidance work together to determine the course to the target, and controlling these functions is the key to this entire discipline. Because of the close interrelationship of these topics, GN&C are combined together as opposed to three separate subjects. Before detailed discussion on the topics, there are a few overarching observations in regards to GN&C.

The broad area of ARL's GN&C research and technology needs further review because what the panel saw was only a slice of all the work being performed in this area. It was not apparent how the GN&C research presented was selected and, further, how it ties to the overall strategy and objectives of the various campaigns. Several researchers expressed frustration about obtaining commercial off-the-shelf (COTS) software in a timely manner. Their comments indicated that Army information technology security does not recognize the mission of ARL and the importance to procure software in a timely manner; that is, ARL researchers want their software needs to have a higher priority. This restriction requires attention and needs to be addressed by ARL management. The procurement of storage for data was also noted to be a major shortfall requiring ARL management action.

The ARL GN&C management indicated that ARL interactions with the Armament Research, Development and Engineering Center (ARDEC) are very healthy, but their interactions with the AMRDEC are fair at best. While the planning process for research project selection in the GN&C area appeared to be very thorough, the presentations to the panel did not reflect that high-level planning process. As a consequence, the presentations did not seem cohesive or to be well coordinated. Finally, through questioning of the oral and poster presenters, repeated lack of in-depth knowledge of the state of the art in GN&C was noted, and there was a lack of references to external experts (elsewhere in DOD, industry, universities) as well as both internal and outside literature.

Kinetic Lethality

The kinetic lethality topic investigates the relationships that cover the flight maneuverability, target acquisition, and delivery to within a designated CEP (central error probable) for guided projectiles and missiles. The goal is to look at design factors proactively in simulation concurrently with technology development that many times takes place at one of the Research, Development and Engineering Centers (RDECs). The milestones for this ARL work are defined, seem appropriate, and appear feasible in the time frames allotted. This 6.2-funded work (early applied research) may lead to shorter-term applications than the 6.1-funded work (basic research), although both may have longer-term implications. Simulation with its verification and validation (V&V) appears to be a cost-effective avenue for ARL. Trying to visualize what will and can be achieved in the long term is a challenging nonlinear feedback problem worthy of increased ARL effort. The work seems to be a good positive progression in this direction. The equipment and tools that are being used are appropriate (but old) for these early investigations. The ARL team collaborates quite often with ARDEC and occasionally with AMRDEC.

The ARL researchers would benefit from greater access and participation within the DOD laboratory system and contractor communities who have, in several notable cases, made advances that may accelerate the research at ARL. Some of the work reviewed has been published in technical journals and conference proceedings, some of which are classified. Exposure of ARL's work is necessary and is achieved through presentations at important conferences. Travel restrictions have been eased over the past 2 years, and funding for travel is being budgeted.

Shock Energy on Target

This R&D project uses a reflective screen and laser illumination to image the reflection of blast wave from a target. The goal is to be able to measure shock wave reflection processes by obtaining incident and reflected shock wave velocities from optical images obtained with a shadowgraph-like method. While an interesting notion, relating shock wave velocities to energy transfer requires the interpretation of complex three-dimensional flow fields in order to learn about the actual interaction of the blast wave with the target surface. Mechanical measurements of the surface deformation (strain gauges, digital image correlation, and force transducers) would directly yield information on the target energy absorption and provide important checks on the optical methods, which will be difficult to interpret except for the very simplest targets.

Impulsive Control for Highly Maneuverable, Small Diameter Munitions

Scoping study on the use of thrust vector control to maneuver Mach 2 gun-launched projectiles and detailed examination of specific aerodynamic issues was presented to the panel. This scoping study used engineering correlations for solid rocket motor performance and conventional vehicle performance, stability and control model based on steady aerodynamic coefficients (lift, drag, moments, and derivatives). Aerodynamic coefficients were based on computational fluid dynamics using steady-flow solutions based on Reynolds Averaged Navier-Stokes (RANS) modeling for turbulent flow. Open loop vehicle performance simulations with elementary rocket motor thrust models demonstrated the ability to change vehicle heading and 150 m offset in cross-range trajectory during 500 ms burn. But this simulation involves very high angles of attack that excite oscillations in the projectile attitude, which results in large-amplitude, slowly decaying oscillations in angle of attack. It is not clear if these concepts can or will be translated into weapons systems even in the long term and needs to be explored strategically to determine if this line of investigation is warranted.

High-G COTS Component Survivability

In an effort to allow for cheaper and faster production of military technology, ARL is looking into the use of COTS components in vision-based guidance systems. ARL has designed experiments to determine and quantify failure criteria of COTS servos with the use of the air guns to reproduce multiple millisecond high-gravity (*g*) environments. A predictive model analyzing probability of failure of COTS components based on changes in duration and *g*-level was determined and later found to be inaccurate due to a lack in experimental severity.

3D-printed materials being used to develop new technologies for the warfighter are being implemented in designs under the assumption that bulk mechanical properties and isotropy apply. Experimental strength properties of 3D-printed glass-filled nylon were found to be largely over estimated in the datasheet given by the material supplier. Internal and interfacial properties are being determined, and soon the strength losses will be explained. These material models are now being inserted into explicit finite element models to further support vision-based component and assembly design.

Scientists and engineers focusing mainly in the area of the sciences for lethality and protection are teaming up with other campaigns with specialties in the area of materials research to combine new materials technologies with scalable dynamic experimental processes in order to determine material susceptibility to high-*g* environments. The findings will determine if these novel processes and materials can be implemented in guidance systems and munitions in order to reach desired lethal effects at standoff ranges in visually constrained environments.

Vortex-Fin Interaction Predictions

Canards are effective aerodynamic maneuver control actuators for guided missiles and projectiles. However, vortices emanating from the canard tips (and sometimes root area) usually interact with aft control or stability surfaces to cause adverse roll moments that can lead to reduced, or total, loss of roll control. The vortex-fin interaction problem has been addressed for decades, but accurate, a priori prediction of the interaction dynamics has not been demonstrated. Engineering-level prediction codes with added vortex modeling reasonably predict the aerodynamic loading (forces and moments) for low-to-moderate angle of attack. Computational fluid dynamics (CFD) predictions provide more underlying physics to higher angles of attack. However, any validation of these simulations is usually restricted to comparing the total loads on the whole flight vehicle, usually without any flow visualization or component (fin/canard) loading measurements. This reduces the ability to determine how accurately the vortex-fin interaction is being predicted—for example, are differences due to the vortex-fin interaction or inaccuracies elsewhere in the flow field? A recent experimental wind tunnel study by SNL provides not only fin loading data, but particle image velocimetry data providing the flow velocity components and turbulence statistics. Validating CFD with this experimental data will allow determination of metrics such as mesh cell density required for adequate vortex resolution, quantifying numerical diffusion of the vortex, and the level of vortex resolution required for adequate determination of the vortex-fin interaction forces and moments. ARL has long been a leader in the use of CFD on high-performance computer systems to study the aerodynamics and flow phenomenology of Army munitions. Collaborations could help to leverage this work to obtain additional validation data and provide a conduit to transfer knowledge gained to the design and development community through improvements to the engineering-level codes.

Validation of numerical simulations (RANS) of vortex-fin interactions against SNL wind tunnel data (M = 0.8) is positive. Strong participation in the American Institute of Aeronautics and Astronautics meetings and DOD/DOE Technical Cooperation Program (TTCP) activities is encouraged. Exploring the transition from CFD++ to new software (Kestrel developed by the National Aeronautics and Space Administration) that will use next generation of modeling (large-eddy simulation or LES), which will enable higher-fidelity modeling of complex flowfields, is also encouraged. Kestrel will need to be extended to include models of control systems and 6 degree-of-freedom rigid-body-dynamics (RBD) that are already included.

Vortex-Fin Interactions on Bodies of Revolution

Accurate determination of flight behavior is critical to the development of new, affordable precision munitions. The current focus for maneuverability is the use of moveable lifting surfaces on the fore-body (canards) of a fin-stabilized projectile. The vortices shed by the trailing edge of the canards are known to have an impact on aerodynamic performance and have been studied for years on high-fineness-ratio missiles. The current research focuses on improving understanding of the canard-trailing vortex on the aerodynamic performance of a short length-to-diameter, fin-stabilized munition when the canards are deflected for a roll, pitch, or yaw maneuver. The in-house research utilizes high-fidelity CFD (traditional and coupled with rigid body dynamics simulations) to increase the understanding of the flow phenomena. Some experimental data for validation of forces and moments has also been collected in-house, and collaboration partners are investigating and collecting additional experimental methods. The increased understanding of the flow phenomena will allow lower-fidelity tools to be enhanced and may eventually allow for the impact of the interference effects to be minimized, thereby allowing greater maneuver authority.

Coupled CFD-RBD simulations of bodies of revolution with fin-canard configurations using commercial software (CFD++) and RANS modeling (the same approach as DeSpirito) is a credible approach. Key issues are related to loss of control authority by fins due to impingements of vortices from canard tips and canard-body joint, particularly when the projectile is at a high angle of attack. Control surface influence can be canceled out or even reversed due to changes in surface pressure distribution and separated flow that result from vortex impingement. Control authority (in terms of blade section aerodynamic coefficients) is a sensitive function of the projectile angle of attack as well as the local fin or canard section angle of attack. The simulations demonstrate the importance of validation against experimental data for specific configurations. Experiments are in progress through informal collaborations with the Air Force Research Laboratory (AFRL) water channel experiments (12-scale model) and visualization experiments at Stanford University and West Point.

Polar Coded Apertures for Compression Spectral Imaging

In recent years, a number of low-cost precision munitions have been developed. These munitions rely on GPS for navigation, which is vulnerable to jamming and suffers from target location error. Vision-based navigation overcomes these deficiencies and has been successfully used in many unmanned aerial vehicle applications. Unfortunately, the infrared (IR) cameras typically employed in military applications are too expensive for low-cost munitions programs. Compressive sensing (CS) allows high-resolution images to be obtained from a low-resolution imager, or even a single sensor, thereby reducing cost. The advantages of a CS IR imager can be directly extended to spectral imaging, which has been shown to increase targeting and detection performance. Initial research focuses on the design of a stand-alone compressive spectral imager in a controlled laboratory environment. The aperture code is optimized both in its geometry and binary pattern to enhance image quality. A continuous motion model is integrated into the CS algorithm to remove image blur. Simulations show accurate spatial and spectral reconstructions using only a fraction of the full amount of measurements. Hyperspectral simulations of up to 128 bands are included. A laboratory experiment is currently in progress to validate these simulation results.

Line-of-Sight Rate Estimation Algorithms

One of the key campaign initiatives (KCIs) is desired lethal effects at standoff ranges in constrained environments. Inside this initiative lies the goal of being able to engage a moving target with a gun-launched projectile. The line-of-sight rate is a component of angular velocity and would be measured by a gyro attached to a gimballed seeker that was perfectly pointed at the target. The gun-launched projectile environment prohibits perfect gimballed seekers, and so the line-of-sight rates need to be estimated from strap-down seekers and subtactical-grade microelectromechanical systems inertial sensors. While others in the missile community have worked with strap-down seekers, prior literature treated the line-of-sight rate estimation problem separately while assuming error-free altitude estimates. Basic research was performed to create algorithms that accounted for nonnegligible altitude errors. It was also found that the altitude and line-of-sight algorithms could be used to aid the seeker-target-tracking algorithms. These algorithms are being implemented onboard microprocessors for a series of flight experiments using real maneuverable projectiles. Estimator performance is being measured against post-processed sensor data that serves as a truth source.

Emerging COTS Measurements and Flight Experiments

Image-based navigation uses computer vision to estimate navigation states of projectiles and provides ARL with a strategy to achieve guided lethality in GPS-denied environments. Extracting navigation states from computer vision results (scene features or target identification) requires modeling the projective transformation performed by the imager system in converting the external scene into a two-dimensional digital image. Errors in the imager model used by the projectile guidance system were seen to result in corresponding errors in navigation state estimates. These errors need to be modeled in order to effectively design the projectile guidance system and predict its performance.

Not only do guidance systems for gun-launched projectiles need to survive the setback loads of launch, but changes to the performance of electromechanical components due to shock need to be understood such that the system design can account for shock effects. COTS imager systems provide ready access to rapidly progressing imager technology, but these imagers are not adequately characterized by the manufacturer and have not been designed for performance in a high-*g* environment. Assessment of the effects of shock on COTS imagers is an area of research not addressed by the academic community studying vision-based navigation for robot applications.

This research effort at ARL presents a strategy for evaluating the performance of navigation imagers in the areas of resolution, sharpness, color perception, projective transformation, and imager boresight. Imagers are characterized both before and after exposure to a shock-event-simulating gun launch. Changes to the imager as a result of the shock event are modeled as a source of error in the projectile navigation states, and this model is incorporated into the simulation environment for design and evaluation of the projectile guidance system.

Summary of Accomplishments

The ARL aspires to be the nation's premier laboratory for land forces. The lethality and protection program at ARL has made significant headway in contributing toward achievement of this goal. They have succeeded in establishing relationships with top RDECs and some university laboratories that are also leaders in these areas—although there are not many. However, ARL needs to broaden its technical network to include other services (some of which may be international) and with major industries in the GN&C research and engineering area. For example, the largest missile company in the world has no one involved (collaborating) with ARL in the areas reviewed. Given ARL's published S&T campaign plans to decide to lead, collaborate, or watch in all technical areas, ARL in the GN&C area needs to decide how important the state of the art in the aspects of GN&C is to the Army and where ARL will engage and at what level and with whom.

Although not immediately apparent, the ARL research team has made significant progress toward developing the technical underpinnings of advanced guided munitions in the areas of aerodynamics, guidance and control, and terminal homing. This includes insights into vector thrust aerodynamics, image correlation, and navigation fixes without GPS, in addition to the potential advances in energetic materials. The work, taken together, appears to define a new generation of precision guided munitions.

ARL has attracted some outstanding personnel, especially new Ph.D.'s, and is undertaking interesting and relevant work on par with academic departments. ARL needs to continue to invest in its staffing in this manner.

Impressive research is ongoing in the areas of vortex-fin interactions research and high-maneuverable, small-diameter munitions research. While potential achievements in these areas are still in progress, several were considered noteworthy. Progress in the turbulence control of vector thrust control seems exceptional for furthering the capabilities of next-generation munitions. Areas such as navigation, image recognition, and terminal homing will likely become a breakthrough when taken together in the context of technology demonstration. Specific examples are discussed below.

Flight Control

This work was well represented by a poster on the topic of vortex-fin interaction predictions of elementary configurations with experimental validation. The principal investigator is well known in this small technical community and very well respected. The caliber of the work is high. Flight control is of course key to smart projectile development, and several topics of similar nature are being pursued. The ARL work is unique and did not overlap with other work, although these topics are standard fare for missile developers.

Navigation

Navigating in a so-called denied environment leads to some reinventing and improvement on some old techniques. The presentation to the panel on polarized skylight navigation was very interesting and well prepared. It introduced work well performed by researchers below the Ph.D. level. This kind of work is important and essential for highly effective R&D organizations.

Shared Vision

The ARL GN&C research team seemed to have a shared vision. However, it was difficult to surmise this vision, as it was not explicitly stated in any presentations. Rather, the vision that emerged from numerous discussions was to develop a gun-launched, rocket-propelled, guided projectile of moderate range. Everyone appeared to either be researching critical elements and features of this vision or performing what might more accurately be called risk reduction for potential technology demonstration (e.g., shock tests). The team has made progress in guidance and control (that the team calls navigation), aerodynamics, and terminal homing parts of the kill chain. However, because the vision is not quantitative, it is not possible to perceive how far the research technology needs to be "stretched" or whether the selected area of investigation is capable of meeting the technical needs. For example, an analysis of the munition concept might show image compression if the image data is not a productive avenue. Alternatively, it is likely that with a short time to go after target detection by the munition, a direct image and comparison with targeting image will be required, with potentially different research issues.

Transition

In discussing how to transition research results to the 6.3-funded research community, the concept of technical demonstrations was mentioned and is a good idea, particularly if performed in a collaborative manner with industry. However, there needs to be a systems engineering context to facilitate the potential transition. This is not in opposition to the spirit of research but can inform the degree of understanding of potential risks as well as critical technologies and phenomena.

Challenges and Opportunities

ARL has a strong legacy of outstanding work in exterior ballistics and maintains a modest core competency today, with significant limitations due to the emphasis on simulation versus testing and experimentation. The limitations of simulation and challenges to progress were clearly recognized in several areas, such as high-angle of attack flow simulation, shock boundary layer interactions, vortex-fin interactions, and rocket motor exhaust with external flow field. The staff are engaged with other laboratories to obtain experimental data sets to validate simulations and working with DOD TTCP programs to exchange information. One of the limitations is the lack of coupling to in-house testing programs (which were historically a strength of ARL) that can be used to both provide validation data for the simulations and to test concepts that are beyond the capabilities of current simulation technology. The effort is an important research area for ARL, but the CFD effort has to be coupled to an equally strong laboratory and a robust and well-funded field-testing effort. This is the main weakness of this program—an excessive reliance on CFD. Lack of in-house testing and a laboratory experimentation program is a definite weakness in this area for ARL. The staff has been proactive about seeking out and piggy-backing on test programs at other DOD laboratories, but without a strong testing program closely coupled to the CFD efforts, ARL will not be a leader in this area. This is a negative for attracting and retaining staff in this area because of the inability to interact with experimenters on a daily basis. To have a robust research group in this area, there needs to be a critical number of experimenters, analysts, and numerical simulation subject-matter experts.

The ARL research community appears to be aligned in their vision of developing and improving the precision of guided munitions. However, it was not clear which performance attributes present the greatest technical challenge to ARL scientists. A number of researchers cited prior open literature and appeared to openly publish in the research areas; however, there appeared to be little awareness of prior DOD sensitive and classified research, development, and even deployed capabilities. For example, approaches for tracking targets moving behind obstacles or through jamming strobes is well described and have been implemented in the air defense and strike communities of all services. As another example, a thoughtful research effort in polarized skylight for munition attitude and guidance (navigation) was presented that might benefit from collaborative engagement with the broader DOD navigation community, even though this community is working in different performance regimes. A Defense Technical Information Center search would likely provide indications of these prior accomplishments and ongoing work and would provide a basis to visit and establish collaborations with the key experts from across DOD.

ARL staff in the flight GN&C area needs to survey prior research (classified and unclassified) and undertake field visits to industry and academia as relevant to select topical areas to increase knowledge of the state of the art in the flight GN&C area. This review needs to particularly focus on the state of the art in the following: (1) target estimation and tracking that has been addressed by the air defense radar community across DOD and airborne surveillance ground tracking technology; (2) gun shock tolerant electronics and actuators that are well characterized in industry; and (3) tracking under measurement constraints, which has been addressed by the radar community. In this regard, the polarized skylight navigation research presented will benefit from broader engagement with the celestial navigation community.

Working to improve accuracy of hypersonic projectiles in adverse environments is a clear challenge worthy of continued effort. An opportunity exists for ARL to improve in GN&C by working more closely with both the other services and with major industry. The ARL work made little reference to work performed by research groups in the other services (e.g., AFRL) and showed almost zero knowledge of work done in industry. In fact, there appeared to be little or no joint collaboration with the largest companies in the United States who have worked GN&C for many years.

Few researchers at ARL have ever used statistically designed experiments or shown use of Bayesian techniques in the analysis of data, even though the laboratory is well known for its work in Dempster-Shafer techniques in other areas of research. This lack is evident in the flight, guidance, navigation, and control project planning, setting objectives, and execution. Improvements that will result in major cost savings and accuracy of interpretation of experiment results can be found by using the design of experiments that has been shown and well demonstrated by the National Nuclear Security Administration laboratories. Also, finding interaction amongst causal variables in determining response equations for experimental results can only be found using design of experiments techniques (e.g., Design and Analysis of Experiments by Douglas Montgomery 8th edition, 2013). Design of experiments remain a critical way to linking cross-cutting technologies and analysis techniques while maximizing optimization of resources and saving costs.

OVERALL QUALITY OF THE WORK

ARL's research on lethality and protection ranges from basic research that improves basic understanding of scientific phenomena to the generation of technology that supports the following: (1) battlefield injury mechanisms, human response to threats, and human protective equipment; (2) directed energy programs; (3) ballistics and blast programs that address weapon–target interactions and armor and adaptive protection developments; (4) disruptive energetics; (5) kinetic lethality—propulsion and launch/effects on target; and (6) flight, guidance, navigation, and control.

Its research on battlefield injury mechanisms is important for ARL because a better understanding of these mechanisms is vital to improving protective equipment. This is especially true for protection of the head, where there is considerable uncertainty about allowable levels of shock, which greatly affects protective options. The most impressive accomplishment of the battlefield mechanisms–human response–human protective equipment program is that a highly competent cadre of scientists is at work and a credible program is under way. A long-term vision for the battlefield injury mechanisms projects could serve as a philosophy that helps allocate resources and set program direction. Almost all of the topics presented in this subsection—battlefield mechanisms, human response, and human protective equipment—had a combination of computational and experimental approaches. The real-time interplay of experiment and computation is needed.

ARL's campaign plans categorize directed energy (DE) as a focused area under the much broader category of electronic warfare (EW), in accordance with the Army's definitions. The ARL posture designations for both radio frequency-DE and laser-DE are to *collaborate* rather than *lead*. The subsuming of DE under EW and a collaborate-only posture indicate that ARL has downgraded the priority of DE within its technology portfolio from its previous robust effort. The consequence of this status change was evident in the current programs presented: they appear to be a small collection of seemingly unrelated projects. ARL needs to take a strategic look at the area of DE to determine its ongoing priority and focus the effort accordingly, with a view to the 2035 time frame; the strategic review needs to include consideration of future capabilities that the Army will need that DE might fill, and what DE capabilities might be fielded by our adversaries for which the Army will need countermeasures. A focused ARL DE program would benefit from a systems-level study addressing future Army missions in which DE could play a role and in which DE effectiveness and alternatives to DE are traded off. A highlight of the overall program in DE is the project on adaptive and scalable high-power phase-locked fiber laser arrays. This work is a notable achievement, is recognized as such by the technical community, and appears to be ready for the next step in transition toward field deployment.

ARL has a strong record of achievement in the basic and applied sciences and engineering of penetration and protection. The R&D described in the armor and adaptive protection area showed how ARL is building on its tradition of excellence to provide the knowledge basis for current and future Army needs in protecting our warfighters. This remains a core competency that underlies Army capabilities across the entire DOD, and it needs to be preserved and nurtured. There was significant evidence of teamwork and integration among the projects in, for example, adaptive protection. Examples of the link of experiments and computational modeling to provide physical insight into problems were especially noteworthy, with potential to aid in developing new designs and exploring new concepts. Benchmarking simulations with experiments and the emphasis on bringing advanced technology (particularly in the X-ray region) to bear on diagnostics were impressive. Developing a predictive capability for damage and fracture in metals, ceramics, and polymers underlies the efficient development of new material systems for protection and for developing approaches to needed penetration capabilities. At present, there is no framework that has this capability. However, experimental, theoretical, and computational advances being developed in other countries are making such a capability seem possible in the not-too-distant future. A systematic approach based on understanding the key physical processes is needed because of the wide range of material systems that are becoming available. Material modeling for these systems would beneficially include reliable modeling of the effects of temperature and pressure—two effects that are mostly underrepresented in much of the computational and experimental effort.

ARL's synthesis effort is a commendable and relatively new effort at ARL to develop a chemical synthesis effort and is encouraged to continue and grow in the future. A blended focus on various applications (propellants and explosives) is needed, rather than just explosives. This is a high–risk/high-payoff effort, so ARL could expect that most candidate materials may not, ultimately, transition to Army applications and systems. The propellant simulation R&D clearly remains a traditional strength of ARL that is positively impacting and supporting warfighter and Army needs, and it needs to be supported and nurtured. Evidence of good interactions between experimental and modeling efforts were seen in the embedded wire propellant project. In the extended solids focus area, it is commendable that ARL has scaled some materials to more significant quantities (grams), and ARL needs to pursue scaling to larger quantities for testing as possible. ARL's multiscale modeling efforts are tackling hard problems like explosive initiation and impact thermal runaway or, specifically, explosive failure. Each of these requires the prediction of rapid dynamic phenomena operating in highly unstable threshold regimes, where subgrid material structure, such as porosity and multipoint material correlations, dominate the transient response. ARL is making good strides in this area and needs to continue to broaden its efforts. A transformative capability that predicts dynamic response on component scales of interest via comprehensive treatment of the scale relevant physics will truly lead to an improved understanding of the response of new and traditional energetic materials.

ARL needs to complement the experimental efforts to date in its energetics and propellant projects with modeling efforts that might suggest alternate geometries (e.g., cylindrical) and perhaps allow additional information (e.g., model parameters) to be obtained from the data. ARL's modeling of solid propellants appears to be state of the art, with the reaction modeling approach using laminate/mixture theory. Further, positive progress was evident in understanding the detrimental formation of cracks and the sudden nonlinear increase in burn rate as a function of pressure. Validation of ARL's modeling effort is clearly ongoing, such as experiments with a wire or foil insert, and is needed as an important component in all energetic and propellant projects. As it is well known that energy release rate is dependent on surface area of the burning propellant, the goal to fabricate an optimal shape propellant using 3D additive manufacturing techniques needs to be pursued. ARL's modeling in this regard to develop appropriate geometric web designs to tailor the burn rate to achieve the ideal constant breach pressure response is encouraged. Additionally, the modeling includes thermal effects (coefficient of thermal expansion), since the burn rate is a function of initial propellant temperature and appears promising. As part of this technical effort, the coefficients of thermal expansion were determined for several propellants and celluloids. Part of the modeling and design processes needs to address the question of whether temperature-sensitive glues could occlude surface area at higher temperatures. Developing 3D additive manufacturing techniques applied to solid propellants and temperature-sensitive glues to control surface area for Army applications is exciting and necessary.

ARL's research team has made significant progress toward developing the technical underpinnings of advanced guided munitions in the areas of aerodynamics, guidance and control, and terminal homing. This includes insights into vector thrust aerodynamics, image correlation, navigation fixes without GPS, in addition to the potential advances in the energetic materials. The work, taken together, appears to define a new generation of precision-guided munitions.

ARL has attracted some outstanding personnel, especially new Ph.D.'s, and they are undertaking interesting and relevant work on par with academic departments. ARL needs to continue to invest in its staffing in this manner. The research in the areas of vortex-fin interactions research and highly maneuverable, small-diameter munitions research is impressive. Because flight control is, of course, key to smart projectile development, ARL's research programs in this area are a positive development and growth areas for ARL and the Army. In the GN&C R&D area, ARL needs to decide in which areas it will lead, versus collaborate, versus watch or follow. Strategic thinking by ARL in the GN&C area—to define how important the state of the art in the aspects of GN&C is to the Army and where will ARL engage and at what level and with whom—is deemed to be critical.

4

Information Sciences

The Panel on Information Science at the Army Research Laboratory (ARL) was charged with reviewing ARL research in the broad areas of computational sciences, information sciences, and atmospheric sciences. A 2-year cycle of review has been adopted for this purpose with the focus in 2015 on reviewing activities in computational sciences and on that portion of the Information Sciences Campaign related to work in system intelligence and intelligent systems (SIIS) and in sensing. The review in 2016 was focused on research related to networks and communications, cybersecurity, human information interaction, and atmospheric sciences.

ARL research in information sciences is focused on developing and enhancing intelligent systems for the analysis of information and knowledge. Included in this approach are technological advances that support information acquisition, reasoning with such information, and activities, including collaborative communications, that support decision-making. Research in these areas falling under the broad categories of SIIS and sensing were reviewed in June 17-19, 2015, at Adelphi, Maryland. The research program in atmospheric sciences includes projects that belong to these categories and were reviewed on June 1-2, 2016, at the ARL Laboratory at the White Sands Missile Range in New Mexico. Additionally, research related to networks and communications (NC), human and information interaction (HII), and cybersecurity were reviewed on July 19-21, 2016, in Aberdeen, Maryland.

Research projects in the NC area concentrate on understanding and exploiting the interactions between information and socio-technical networks—in particular, on communications and command and control networks. Areas of specific endeavor include channels and protocols, network control and behavior, and network-based information processing. The research is primarily aligned with the ARL key campaign initiative (KCI) of taming the flash-floods of networked battlefield information, which aims to realize analytical approaches that better describe, characterize, and explain complex, dynamic, multi-genre networks and the data generated by these networks. The research is also aligned with ARL core campaign enabler (CCE) of networking and communications in contested and austere environments.

The Army faces significant and growing challenges in the area of human and information interaction. The growth in use of unmanned aerial vehicles, robots, bots, and social media is creating a challenging communication environment for the Army of 2040 in which humans and machines communicate, but often without the ability to determine whether the communication is from and to a human or a software agent. There is a need to understand how to communicate effectively in this environment and to accurately understand, and in some cases predict, the communications of other actors, so as to make timely and accurate decisions. The need to operate effectively in this environment is creating a need for a new transdisciplinary science of human and information interaction. Human and information interaction is central to the third offset strategy[1]—particularly the areas of human-machine teaming, cyber warfare, and the development of new operating concepts.

Research in cybersecurity is focused around theoretical advances and model development related to cyber-threat detection, recognition, and defeat mechanisms. The emphasis is on both detection and defeat of highly sophisticated attacks that use techniques very different from those encountered in commercial or civilian settings. The combination of strategic and tactical networks in use by the Army creates unique challenges in the domain of cybersecurity. The dual nature of ARL's cybersecurity role, operations, and research create a truly unique capability for cybersecurity research. ARL is one of the few cybersecurity research organizations with continuing access to real-world data, the importance of which cannot be overstated. Additionally, ARL researchers interface with a deployed operational environment. This gives ARL an ability to focus on research that actually matters. Taken together, real-world data and access to operators positions ARL's cybersecurity research to be exceptional. Given that it addresses a constantly evolving environment of threats, cybersecurity research tends to be more applied than pure theoretical advances; the ARL portfolio of research in this area contains an even balance between the theoretical and applied components.

The Army faces significant challenges in both the characterization of the battlespace environment and prediction of optimal conditions for engaging an adversary with overwhelming force. The Army is likely to find itself engaged in a number of challenging environments, from complex terrain to sprawling urban areas. There is a need, therefore, to characterize these diverse environments and develop accurate, relevant, and timely predictions of their future state on spatial and temporal scales useful to Army operations. The related need to collect and process accurate, relevant, and timely environmental characterizations in austere conditions, and translate that data into actionable environmental intelligence for field commanders, will also pose new challenges to the computational sciences community. The research being conducted by the Battlefield Environments Division is addressing these challenges and includes a mix of analytical, computational, and experimental projects in its atmospheric sciences program.

SYSTEM INTELLIGENCE AND INTELLIGENT SYSTEMS

Research projects in SIIS were presented in three thematic areas: information understanding, information fusion, and computational intelligence. Collectively the work addresses technical challenges in the use of sensors, communication, and computing to provide the soldier with new levels of tactical intelligence and the automated support needed for missions. Areas of specific endeavor include language translation, information extraction, semantic analysis, understanding of human trust networks, fusion

[1] This is a Department of Defense initiative. A core component of this initiative is the formation of a Long-Range Research and Development Planning Program that will target several promising technology areas, including robotics and system autonomy, miniaturization, big data, and advanced manufacturing, while also seeking to improve the U.S. military's collaboration with innovative private sector enterprises.

of conflicting information, integration of video and text analytics, anomaly detection, reasoning under uncertainty, robotic control and path planning, and models of cognition and tactical decision making.

Accomplishments

The research portfolio contains an appropriate mix of theory, computation, and experiments and is clearly of value for the support of Army missions. The research projects were generally, but not uniformly, of a good quality. The strongest work has been published in elite conferences and archive literature, and additional opportunities exist to disseminate other work at respected venues. It was difficult to identify a unified theme or common basis for the multiple projects, and the principal challenge for each of the three areas of research is to fuse the collection of projects into a coherent whole.

Information Understanding

Research in this theme focuses on the development of critical methods and techniques for transforming data so as to provide useful information to the soldier. The work represented solid but incremental advancement in several interesting and potentially important areas such as language translation, information extraction, semantic analysis, and understanding of human trust networks. Some of the projects appear to have potential for short-term applications in the field. The research staff ranges from experienced researchers with doctoral degrees to students pursuing master's degrees or internships. Laboratory resources appeared to be adequate to support the research agenda.

The work on temporal information extraction focuses on methods for extracting temporal relationships from text for constructing knowledge networks. The proposed approach is technically solid and particularly well grounded in the available literature. Temporal relations are an important area for future work by search engine companies, and the long-term applications to the Army are clear. This work has been published in a top conference—Association for Computational Linguistics (ACL)—an indication that it is first-rate quality work. The impact of the work is already evident in that other researchers at ARL are making use of this work.

Ongoing research on influence in social networks is closely related to activities in the network science Collaborative Technology Alliance (CTA) and pertains to identifying mechanisms for trust formation in human networks. The effort has emphasized identifying main factors, but the work has not addressed this issue in the context of networks. Most results to date are based on an analysis of data from the online microfinancing company Kiva and other lenders and are related to microfinancing and corruption perception indexes. The opportunities for extending this work to intelligent systems research for the Army were not readily apparent. The project would benefit from additional awareness and understanding of the literature in the area of online trust.

Also stemming from work related to the network science CTA is research related to agent-based semantic analysis in information retrieval. This research focuses on how to manage information dissemination across constrained channels based on the trade-off between size and accuracy. Rather than posing the task as a constrained optimization problem, the proposed approach implements a fuzzy logic model in order to include model uncertainty or fluidity in the attributes. The work requires a clearer articulation of what additional insights into operational questions are available from the proposed methodology. Further, to establish the advantages of the approach, it would be helpful to compare results against those obtained through more traditional optimization techniques. If such optimization techniques have inherent limitations or are not applicable in this particular problem setting, statement of the problem could better articulate those techniques.

An example of a 6.2 (applied)-level research program addressing a real-world need is the project whose objective is to perform language translation in an automated fashion with a goal of enhancing translation capabilities for low-resource languages. The approach involves the construction of a language model based on entropy minimization that selects those sentences that need to be translated by humans in order to construct a best overall training corpus. The novelty of the work centers on methods for constructing the language model.

Information Fusion

Research in this area pertains to the fusion of data from disparate sources to produce timely, actionable information for the soldier. The research has advanced the state of the art in several interesting and potentially important areas such as fusion of conflicting information, integration of video and text analytics, and anomaly detection. The quality of the research was generally good, ranging from archival quality results publishable in top journals to preliminary but promising work. In some instances there was an apparent lack of full awareness of critical literature and alternative approaches. The researchers engaged in the projects were well qualified and possessed a range of experience. While laboratory resources appeared to be adequate to support the research agenda, there were not enough experienced staff to fully develop the ideas.

Research on estimating credibility by fusing subject opinions was of good technical quality and was disseminated in a reputable conference. This research focused on fusing multiple inconsistent and potentially conflicting information sources. The work treated inputs as propositions with truth-values and used subjective logic in a manner that took into account prior experience with the sources. It was well grounded in statistical machine learning approaches, but it could benefit from consideration of prior work in the human computation literature.

A project that seeks to develop an observer model for helping soldiers determine salient targets draws upon recent advances in image processing and neuroscience. The approach is to train a saliency model on the basis of experiments with experienced soldiers and then adapt that model for automated training of less experienced soldiers. The work emphasizes anomaly detection in saliency models. It is well grounded in previously published work and could be advanced by research considering information from a network of devices.

In a related vein, research is also being conducted into integrating complementary or contradictory information into a fuzzy model in order to determine the value of the information. The underlying idea is to take multidimensional values of information metrics and to project them onto a single dimension. This kind of dimensionality reduction facilitates subsequent sequence-based analyses. It is not clear that such analyses could be performed without unnecessary loss of information. This work would benefit from considering the results of previous research in indexing, feature reduction, and information theory.

Several research efforts in the thematic area of information fusion were in the early stages of investigation. An effort dealing with intelligent information management in the battle environment focused on bandwidth management during distribution of information to end users. The approach is based on linear scoring functions with weights established by the users. Its effectiveness is based on questionable assumptions about human capability, and the investigators might want to consider alternative approaches in the literature that deal with learning user preference and modeling utility in humans.

Another early research effort that could potentially have important impact is based on a hypothesis that learning from video and text could benefit from being done together when video includes or is associated with text. The improvement in the quality of the results comes with the cost of implementing joint learning methods, a much harder problem to solve. The next appropriate steps would include a literature review that examines relevant advances in optical character recognition, video retrieval, and object recognition in videos.

Computational Intelligence

Research in computational intelligence examines the development of intelligence in systems to support highly automated or autonomous operations in support of Army missions. The research efforts have yielded significant advances in several interesting and potentially important areas such as reasoning under uncertainty, robotic control and path planning, and models of cognition and tactical decision making. Researchers engaged in the effort are well qualified, but the team comprises a disproportionate number of early-career researchers. Some of the research efforts lack the staff that would be needed to fully develop the ideas. In particular, projects related to information for robot navigation and message delivery and those requiring expertise in psychology would all benefit from strategic recruitment in targeted areas of expertise.

A research project focused on improvement of visual classification for navigation purposes uses an approach of implementing partially supervised discovery and labeling the navigation domain. The research aims to reduce the effort associated with human labeling of terrain and objects in images for the purpose of supervised learning of objects in images. Over-segmenting the image and then using clustering techniques on the resulting segments accomplish this goal. Human labeling based on the clusters is much faster but less accurate. However, this loss of accuracy does not significantly degrade robot navigation when using the new visual system. This research is expected to have an important impact and has yielded promising initial results.

Another promising research project uses a semantic vector space for reasoning in the presence of uncertainties. This work seeks to take advantage of two different types of semantic models with a goal of augmenting a curated knowledge base by reasoning through analogies based on statistical representations. Both the ideas and the proposed methodology contain novel elements, but the work is still in an early phase. A number of complexities have to be resolved, including those arising from multiple meanings for words. The work is well grounded in the literature, and the researchers are aware of related efforts in the research community. With continued support and application to meaningful problems, it has the potential for publication in top journals.

Investigations into the concept of robust distributed communication relays for minimizing message latency in a vulnerable and uncertain operating environment are also ongoing, albeit at a preliminary stage. The research is focused on resource-constrained, dynamic environments, where it may be advantageous to use kinetic or nonelectronic means to augment electronic information distribution. The formulation results in a very hard computational problem based on vehicle routing. The approach to handling the computational difficulty derives from an integrated heuristic/high-power computing (HPC) methodology and is being tested in a simulation environment. While the pertinent literature has been explored and cited, additional benefits in computational efficiency may be gained by using recent work on probabilistic spanning trees.

Another example of an early research project investigates strategies employed by humans in situations similar to those experienced by soldiers facing trade-offs between short-term and long-term actions. Initial results in two-player games, which have been obtained using computer agents, demonstrate that there is no dominant strategy. The underlying hypothesis for the work is that these games will provide an environment for investigating how humans move between strategy, tactics, and actions. Although at a preliminary stage, the work addresses an important problem. It is critical to place this research in the context of the literature on human cognition, training, game theory, and computation intelligence in games. Another effort at modeling human cognition was based on varying the volume and velocity of data presented to shooters who have to make a choice of targets. This allows investigation of the effects of data attributes (in classic big data terms) on human ability to make decisions. This topic appeared to be narrow research toward a master's degree, and its future is unclear.

Ongoing work at a more advanced level looks at the use of an information-gain metric to design paths for autonomous mobile robot movement. Entropy minimization concepts are used to develop optimal routes for mobile ground robots. This work has evolved to an advanced demonstration level; there is a good plan for future work that would make use of recent advances in combinatorial optimization methods and that would include constraints on power and available onboard computational bandwidth.

The concept of episodic memory consolidation and revision, or robotic dreams, was used to identify precursors to events such as the deployment of an improvised explosive device. The research is premised on the hypothesis that this computationally demanding problem is intractable in a real-time computation environment but is more amenable to large-scale background processing using otherwise idle computing resources. While this problem is of significant importance, it was unclear what level of computational efficiency was added through the episodic memory consolidation approach vis-à-vis simply devoting additional computational resources to the problem directly. It was not readily apparent how the approach is distinct from traditional work-stealing methods for background processing during low computational loads.

Challenges and Opportunities

The most significant challenge facing the SIIS program is the development of a coherent and unifying thrust for the research program. Overall, there was little evidence that the whole would be greater than the sum of the parts that were reviewed. The portfolio of research is broad, with many lightly staffed projects. A challenge lies in a better integration of the work to pursue fewer, but more significant and more ambitious projects. Such prioritization is important to focus the limited staff on strategically selected projects so as to more fully develop the ideas. It is possible that the above impression is due to the selection of the topics presented to the panel. ARL leadership noted that projects had been selected to provide a broad sample of the research portfolio rather than to explore its depth. If there are elements that more coherently unify the research projects, then it is important to highlight such integration, not just in the top-level overviews but also in the project descriptions themselves.

Intelligent systems research is increasingly dependent on harnessing vast quantities of data, and the fields of scalable machine learning and big data analytics are advancing rapidly. A number of important techniques and tools have been developed to address these emerging challenges, many in the open source community. Some of the projects reviewed in this report would benefit from understanding and using these new tools. Such an approach is being practiced in research related to cybersecurity and other areas that were not reviewed in 2015.

Some of the reviewed projects were particularly strong because they combined pertinent approaches—for example, the combination of statistical/machine learning (ML) methods with linguistic rules—to address research challenges. Beyond such innovative and opportunistic approaches, some impactful research projects demonstrated clear integration with other activities and research at the ARL and other external entities. Other opportunities for such integration exist and could be more broadly embraced.

New advances in automation have changed the roles of humans and machines in intelligent systems to a degree that humans and machines need to collaborate in task activities. Therefore, the human systems perspective has to be fully integrated into any development to ensure usability and robustness of the results. The ARL presentations did not include the work on human interactions, making it difficult to assess the impact of some of the systems intelligence research.

SENSING AND EFFECTING

Research projects in sensing and effecting covered thematic areas of non-imaging sensors, image understanding, sensor and data fusion, and radar signal processing. There was a strong focus in the area of acoustic sensors that collectively examined new materials applications alongside the development and implementation of better signal processing capabilities. Cross-modal face recognition was an important thrust in the image-understanding research. The use of long-wave infrared polarimetry to facilitate discrimination of manmade and nonmanmade objects was another thrust in this research. The sensor and data fusion focus was largely on approaches for dynamically adapting information to situational changes and appears to comprise engineering advances as opposed to exploring fundamental science. Research in radar signal processing was both relevant and of good technical quality, and it focused on signal processing in congested and cluttered environments.

Accomplishments

The research portfolio contains a mix of projects in areas that are well aligned in support of future Army missions. The research was generally of a good quality but not uniformly so. The research projects emphasized applications, as opposed to the foundational science. Some of the work is being published in top venues, but researchers could be given additional guidance and encouragement to present their work at leading conferences and then pursue publication in top archival journals. In general, the research staff were well qualified and demonstrated a good understanding of the important challenges in their work. The long-term research vision provides a natural framework for integrating the individual efforts. However, the connection between these individual activities and this overall framework was not clear. Finally, the time horizon for some of the projects can be shortened (to 3-7 years) to demonstrate proof of concept; the pace of technology advances in the field dictates this necessity.

Non-Imaging Sensors

Research in this domain is primarily on the application of acoustic sensors for Army-relevant tasks, including identifying helicopters through acoustic signatures, localizing gunfire, and detecting vehicles or weapons at long range. The Army has a long history of using such sensors, but new technologies and new signal-processing techniques could extend the range and the precision of these sensors, reduce complexity, and minimize power requirements.

Research related to the application of new materials in the design of windscreens for acoustic sensors has yielded positive outcomes. Acoustic sensors covering a range of frequencies, from infrasound to ultrasound, were explored in this study. Ongoing efforts are evaluating a range of porous materials and are characterizing their efficacy relative to baseline materials used in traditional microphones (e.g., foam). The work is based on results from partners at the University of Mississippi. The results have been presented at good conferences but have not been submitted for publication in archival journals which needs to be done; internal technical reports document the progress of this work. ARL needs to consider the submission of this work for publication in archival journals. There is an opportunity to use foundational theoretical knowledge to guide in the selection of optimal materials for this application.

In research related to enhancing the effectiveness of acoustic sensors, the work on identification of helicopters through the use of innovative signal-processing techniques has yielded promising results. The signal-processing algorithm takes advantage of the relatively invariant blade speed to determine the Doppler shift and to calculate appropriate motion compensation, which is then used to auto-focus. Papers describing the approach have been published in recognized journals.

Research in this domain is also exploring sensors alternative to the more traditional microphone arrays. In particular, a research project is directed at investigating the use of microelectromechanical systems (MEMS)-based, three-dimensional acoustic particle velocity sensors. A commercially available sensor was adopted for this work, and the ARL research is focused on developing signal-processing techniques better than those provided by the vendor for applications that include localization for small-arms fire and triangulation of continuous waves. These devices have significant potential for reducing complexity, weight, and power requirements. Their robustness in field applications continues to be an area for research and development. Findings of this research have been presented at recognized conferences and documented in internal technical reports.

Image Understanding

The image understanding research program was generally of high quality. The work is being published in high-quality journals and shows a comprehensive understanding of research conducted elsewhere and how the ARL research fits into the broader research landscape. The cross-modal face-recognition work represents an excellent example of an appropriate applied-research topic for ARL, given the mission need for such an approach. The researchers were able to articulate the unique Army needs in these problems and were addressing them.

The research on polarization shows strong potential, and continued collaborations with camera producers would enhance capability. Researchers demonstrated an outstanding ability to summarize the significance and impact of the work. Additionally, the collaboration in this project related to sensor algorithms for polarization imagery was considered to be a positive outcome of the ARL open campus initiative.

The image understanding work was applied research rather than fundamental. Appropriate laboratory facilities were available to carry out this research. The work was judged to be relevant to Army needs and could ultimately be transmitted to the field. In particular, the manmade object discrimination work recently resulted in a patent, and a transition path is under way.

Sensor and Data Fusion

The research efforts in this area tended to be more incremental engineering advancements, as opposed to addressing fundamental questions. As an example, design concepts for dynamically adapting information to the situational changes in the utility of data attributes (e.g., accuracy, latency, reliability, data rate) could be a fundamental topic for investigation; instead, it was treated as a human factors problem rather than as a broader concept for automated management of data prioritization. Given the engineering development focus of much of the work, the researchers did not present the current state of the art in sufficient depth, and they appeared to lack access to the field data that would have allowed them to do so.

Research related to detection of vehicles, personnel, or targets is key to Army operations and requires the use of multiple sensor arrays. The fusing of information from many sensors observing similar objects (dependent data) is key to developing better inference. The research focuses on developing a means to fuse correlated information.

The work on dynamic belief fusion applies a sound and seemingly straightforward approach. It integrates the outputs of different object detectors by assigning ambiguity levels derived from previous performance to each detector. An approach of dynamically allocating probabilities based on prior performance has been developed that demonstrates the improvement in detection accuracy over conventional fusion methods.

Another area of research is aimed at developing and enhancing tools to reduce the time between data gathering and making decisions. While important from a practical perspective, the research effort is simply an extension of a fuzzy logic–based tool to assess value of information (VoI) through the use of additional membership functions. The work is in an early stage, and the fundamental technical advances of the approach are not readily apparent.

Across the entire research endeavor, access to actual field data would further enrich the research effort and would help to distinguish the work from that in the outside research community. In the area of dynamically adapting information to situational changes, for example, research in the commercial applications arena focuses on certain performance objectives that do not meet the requirements in the military context. It is important to build in recognition of ease of disruption and cost of errors (e.g., of human life) as explicit considerations in the performance objectives as field data become a critical component in evaluating such criteria.

The laboratories and infrastructure support were appropriate to support the research; the access to the E/H-Field laboratory is particularly beneficial to ongoing research. The researchers were academically well prepared to undertake the work.

Radar Signal Processing

The research reviewed in this domain was of good technical quality and showed promising results. The research problems were well defined, the methodology was explained at an appropriate level, and the results were well organized. Detecting moving personnel under tree cover on the basis of frequency-modulated continuous-wave (FM/CW) radar is an important challenge. A full-wave approach, realized through the use of a parallel three-dimensional finite-difference time-domain algorithm, is deployed for this purpose. Likewise, the ability to detect targets obscured by artifacts in ultra-wideband imagery is also relevant and important from an Army perspective. Mission requirements dictate effective performance in the presence of uncontrolled radio frequency (RF) transmitters, and RF interference notch-filtering techniques are widely used in such applications. An alternative approach based on sparse representation and recovery of signal was proposed that does not have the shortcomings of notch-filtering techniques.

The work related to nonlinear radar methods is also timely. It focuses on the use of nonlinear harmonic radar to achieve greater sensitivity across a narrow frequency band. These approaches have been in the literature for a few years, and early applications in the field represent an important next step. Based on these experiments, a 3- to 7-year time horizon for transitioning to full-scale developments in this area is a real possibility.

HPC facilities at the ARL are a key enabler for this work, especially in applications to detect moving personnel under cover. In a similar vein, the significant use of radio equipment indicates adequate infrastructure to support the research effort.

Two of the projects employed computer simulations to generate input data that were used to demonstrate improved receiver processing techniques. While adequate for purposes of showing proof of concept, the use of real data would also significantly benefit further studies. Simulated data might lack some of the complexities in real data and could lead to false validation of the proposed methodology. ARL might be able to get such data from other Department of Defense (DOD) laboratories or from the industry. If necessary, simple experimental set-ups could be built to collect such data.

Challenges and Opportunities

The overall technical quality of the research in this area was good, albeit with a greater focus on technology development than on foundational research. Overall, researchers are aware of relevant work in their domain of interest, but this was not always the case. The research problems are important to the Army and are unlikely to be pursued in academic institutions or other government research laboratories.

From an engineering standpoint, the researchers seemed knowledgeable about identifying and adapting or creating new signal-processing techniques to the target domain and good at doing this. From a foundational standpoint, however, it was less clear how well the algorithms or ideas would transfer to other domains, or how knowledge would be gained to better understand the underlying physics of the domains.

In the area of image understanding, an increased emphasis on developing rich data sets and the quantitative improvement of performance for larger data sets—especially data sets that include mission-relevant variability—would enhance the impact of the program. In particular, a better understanding of how easily an adversary could defeat cross-modal recognition efficacy is of interest. The research emphasis could be expanded to embrace the fundamental science in order to better understand the physics that drives cross-modal features and the information-theoretic fundamental performance bounds. This would enhance the overall quality of the research and produce results with longer-term impact. Additionally, ARL could better connect with and provide intellectual leadership to the broader facial-recognition community by creating and curating open, standardized data sets as well as challenge problems for researchers to explore.

In the area of sensor and data fusion, dynamic automatic control of data attributes with operator support represents an important emerging area of research. This is especially true given the explosion in sensor technology and limitations in both communications bandwidth and human–computer interaction. A new opportunity would be to expand the focus to automatic data control rather than limiting control to the operator interface as the only basis for stimulating data control actions. While industry wishes to develop cryptographically solid authentication and authorization, the needs of the Army are more complex. While the authorization to manage parameters of a system resides with a few individuals, provisions have to exist for others to take over in the event of unauthorized access by others, or at least prevent it. One possible area for research is to design systems with redundancy, so that if one sensor net or dashboard is subverted, it would be immediately detected. Similarly, based on knowledge of the system, the ability to identify system behavior as illogical, and possibly compromised, would be hugely beneficial in such field operations.

The radar signal-processing work could benefit if the engineers working in this area had graduate-level training. There have been significant advances and new tools in the radar signal-processing area, and familiarity with these would elevate the research. It would also help to improve the success rate for presentations at leading conferences and publications in top journals. Access to advanced technical knowledge online or by other means can be explored. The work related to nonlinear radar shows promise for immediate transfer to a range of battlefield applications. Finding ways to transition this technology as rapidly as possible would be extremely helpful in meeting critical needs in Army operations.

Wireless (and other) communications is a critical part of Army operations today, and its importance keeps increasing. It would be beneficial for ARL to build some capability in this area, at least to support other activities in the laboratory. Lack of this communications technology skills set will increasingly become a disadvantage.

NETWORKS AND COMMUNICATIONS

Accomplishments

Research within NC addresses the technical challenges of communications in highly dynamic wireless and mobile networking environments populated by hundreds to thousands of networked nodes. The research is generally of high quality and relevant to the needs of the Army. The research portfolio is a good mix of foundational theoretical work, with a focus on applications and technology demonstration. In some projects, experimentation is effectively deployed to both validate theoretical models and to improve theoretical modeling. In general, the researchers demonstrate good awareness of related work in other research communities. For some of the interdisciplinary research projects, however, there is uneven awareness of the literature and accepted approaches across the constituent disciplines. Additional focus on formulating relevant research problem statements would be beneficial for these cross-disciplinary research endeavors. The research results are being published in archival journals and in respected conference proceedings. The ability of researchers to travel to key scientific conferences will have a continued positive impact on the overall research environment.

Channels and Protocols

In the area of channels and protocols, the research challenge is to explore new heterogeneous approaches to information delivery across a communication network using previously unexplored network configurations, channels, protocols, and allocation techniques. Research tasks in low very-high-frequency (VHF) range communications, ultraviolet (UV) communications, and on quantum methods for networked communications were evaluated. Research to evaluate the low very-high-frequency range for networked communications represents an effective applied research effort that focuses on both experimental validation and prototype communication node design. The program also reported on the development and demonstration of very small antennas that enable use of this frequency band with small nodes. While this is an applied research and advanced technology demonstration, it is nevertheless of good quality and highly relevant to Army needs.

Work on UV communications included an effective experimental program to validate theoretical propagation models for long link-distances that previously were not explored experimentally. This experimental program has resulted in improvements to theoretical modeling—a notable example of synergistically leveraging theoretical and experimental research for high impact. Finally, in the quantum area, work related to theoretical analyses on quantum key distribution and understanding of quantum entanglement was evaluated. This research represents sound theoretical contributions to the area. This research effort is modest in size when compared to much larger groups addressing these issues elsewhere, both theoretically and experimentally; nonetheless, the work was assessed to be of good quality.

Research related to the common sensor radio and associated medium access control (MAC) protocols were also reviewed. This is a notable example where a sustained and high-quality research effort has resulted in a significant impact. The program has evolved from an early radio design into a communications and networking system that is well suited to address unique Army and other DOD communication network needs, especially for ground operation in which the propagation loss is very high. This program is seeing multiple transitions to other military agencies and to the industry.

Information Delivery

A significant subset of the information delivery research topics included efforts focused on emulation and simulation tools for experimentation, as well as information theoretic foundations, trust in networks, and decentralized learning.

Several projects are aimed at developing emulation and simulation environments for network science experimentation. Many of these leverage the newly opened Network Science Research Laboratory (NSRL). The NSRL provides an open collaborative space for bringing together teams to address multidisciplinary networking challenges, provides flexible, configurable platforms for conducting experiments and studies, and can be used to provide realistic demonstrations of newly developed approaches. Research efforts in this area include video processing over networks, adaptive information query, distributed computing architectures and network protocols, controlled experimental testbeds to emulate field experiments and exercises, and multi-genre node interoperability in networks. The thrust on development of emulation and simulation environments in a research platform like the NSRL has yielded a powerful capability that will enable significant network and cross-coupled research investigations that address future ARL and Army research needs. It serves as a model of enabling capabilities for multidisciplinary collaborative research for complex problems.

The research on semantic information theory and on decentralized learning is of high quality and represents fundamental research aligned with future information systems needs and KCIs. The research on trust-based methods to address decision making while maintaining network efficiency and security is relevant to the NC research challenges and has been published in some leading venues and has received other external recognition. The significance of this research could be clarified by describing how it enhances the state of the art or relates to prior work, and why the trust models are appropriate to complexities of future battlefield communications.

Research on opinion dynamics would benefit from a stronger consideration of problem statement and prior work.

Control and Behavior

The research projects in control and behavior were primarily theoretical and considered algebraic topological approaches to address network coverage and topological persistence. This work is categorized as foundational, with strong theoretical underpinning, and identified as a core campaign enabler; it is more fundamental in nature and did not appear to directly address a KCI. Also reviewed in this category was work related to network topology control to maintain robust network performance in contested environments. The overall quality of this research is high.

Challenges and Opportunities

For much of the NC research portfolio, it is often unclear how the individual research topics address the scientific challenges of the campaigns—the KCIs and CCEs. The introduction of KCIs and CCEs to bring the research topics into focus is viewed as a positive step, yet in many cases the connections between the campaigns and the individual research projects are not clear. It would be helpful to see how the campaigns lead to a coordinated portfolio of research projects, and also how individual projects integrate to a comprehensive portfolio that successfully addresses the campaign objectives. Top-level overviews and project descriptions need to highlight elements that more coherently integrate the research projects. Continued convergence between the campaigns and the individual research projects would improve the coherence, and thus the impact, of the overall research program.

For some of the cross-disciplinary research projects, there was uneven awareness of the literature and accepted approaches across the constituent disciplines. This was particularly noted for those projects that address human-centric topics such as trust, opinion, and value of information. To ensure high impact and to avoid compromising the overall quality of research, it is important for these cross-disciplinary research endeavors to establish strong, relevant research problem statements. Researchers pursuing such cross-disciplinary projects are encouraged to establish collaborations and strong mentoring relationships with experts in allied areas early in the problem-formulation stage. In areas where ARL has the quality research expertise in its existing workforce, this need can be addressed internally. The KCI structure will drive further cross-disciplinary research, and the need for cross-disciplinary collaboration and mentoring will increase. ARL leadership is encouraged to proactively support early engagement of these connections.

The concepts of value of information and quality of information are helpful ways to develop bridges between the NC research community and the HII community. There does not appear to be a common agreement on what these measures mean and how they are used. A common understanding and agreement on these measures would facilitate fruitful cross-disciplinary research endeavors that consider human effectiveness and network functionality jointly. The topic of resilience of networks was deemed to be of high importance for Army applications. The program and projects did address the issue of resilience at the application level. There was no evidence, however, of an explicit research focus in this area.

Intelligent systems research is increasingly dependent on harnessing vast quantities of data, and the fields of scalable machine learning and big data analytics are advancing rapidly. A number of important techniques and tools have been developed to address these emerging challenges, many in the open source community. Some of the projects reviewed in this report would benefit from understanding and using these new tools. More generally, the NC portfolio needs to include projects that address some of the challenges and opportunities that these vast quantities of data afford to Army application scenarios.

It would be useful if the NC portfolio included a mix of projects—some that are more narrow and focused, emphasizing specific contributions to a very well defined scientific or technical question, and some that are broader and that combine specific new contributions in a system-level effort, to solve a very important problem in an Army specific application.

HUMAN AND INFORMATION INTERACTION

Accomplishments

The Army finds itself today in a completely "informatted" environment, one where information and procedures are constantly updated on multiple cyber devices in use, and where this information can not only be retrieved but also affects or changes the state of the device. The efficiency with which humans can interact with such a deluge of information has given rise to research in the area of human and information interaction. Within ARL, HII is a new endeavor and has not had sufficient time to develop strong areas and lines of research. As characterized by the researchers and ARL leadership, the current research projects represent the following three major thrusts: (1) Dialogue—How can computational technologies help facilitate communications between humans and devices such as smart phones, robots, or even traffic lights, using natural language, utilizing images as appropriate, and in a manner that captures intent, even in a degraded information environment? (2) Immersed decision making—How do individual humans make decisions in a completely informatted cyber environment—where they are immersed in information from the built environment, social media, traditional media, and a myriad of sensors? (3) Social computing—How do groups shape the information environment, gain information dominance, and take actions and affect appropriate response in a fully informatted cyber environment consisting of societies of bots, gangs of coordinated sensors, and crowds of humans—all connected and working in tandem to effect change?

HII is not human–computer interaction, a field that focuses on the one-person-one-machine interaction. In contrast, HII is concerned with groups—groups of people, software agents and machines, and groups themselves. HII is not social media or social media analytics either, because it deals with more than the content of the communication, and more than just understanding, interpreting, and predicting human and bot behavior in a completely digital world.

The area requires a truly interdisciplinary approach involving sociology, anthropology, linguistics, psychology, cognitive science, information science, and computer science. To build meaningful research in HII, ARL will need to leverage expertise of other researchers within the laboratory as well as the broader external scientific community.

HII is new at ARL, and it was not surprising to the panel that projects are in the early stages, of short duration, and with connections to more detailed 3- to 5-year plans. The status of research accomplishments may be classified as early stages in the HII context, but they build on more mature efforts in related areas. There are both opportunities and significant challenges in the selected projects; each has the potential of becoming a showcase project in this area.

A Lexically Informed Event Ontology

This project develops an a sense-making system for events that draws on existing ontologies such as wordnet and entities, relations, and events (ERE). Such a system could play a critical role in supporting communications in the new cyber environment. Access to an ontology expert would greatly benefit the research program; the investigators could consider taking an intensive course on ontology or include an ontology expert as a collaborator on the project. The use of ontology such as ERE is also limiting because it is not in the public domain. This limits the overall research productivity as well as the venues where the research may be published.

Intelligent Information Management for the Battlefield

In this research project, the investigators are using a policy-based approach to determining the value of information in order to reduce information overload. This is a significant project as it represents good science and is research that has potentially important impact. The research focus is transitioning from the technical side to the human interaction side. The challenges are related to bringing in appropriate information visualization capabilities and getting sufficient and appropriate experimental subjects. This could be facilitated by greater collaboration with those in information visualization (possibly in ARL and beyond) and with access to subjects earlier in the design process. Such an approach could be a template for other strong projects in the HII area.

Situated Human Modeling Via 3D Eye-Gaze Information

This project aims to improve the ability of autonomous agents to communicate with humans. The scope of the research was intentionally narrowed to focus on issues of using on-head eye-tracking glasses to allow autonomous agents to build models of humans. This narrowing of focus has enabled progress to be made, but at the expense of limiting advances in the basic science. The key challenge lies in developing deep theory about humans—specifically, a theory of the mind for HII. This could be facilitated by collaboration with cognitive scientists working in this area.

Each of the aforementioned research projects is an example of HII science that could make an impact in the field. In each of these projects, the researchers demonstrated a sophisticated understanding of the problem, were familiar with some of the relevant literature and were able to draw upon it, and could clearly articulate either the problem being addressed or the value of the results. Each of these projects can be shaped to address an issue in at least two of the three main thrusts for the HII area. Having projects that address a combination of thrusts is critical in developing the overall area and the underlying science.

Challenges and Opportunities

There are a number of potentially important theoretical concepts being addressed by investigators in HII, such as trust, value of information, and quality of information. However, it was found that these terms were often used superficially and without the investigators having a deep understanding of the underlying phenomena. Relevant literature was often not cited, and there was a lack of understanding of the breadth of sciences that already use the words. In many cases, the investigators were unaware that the concept had been conceptualized in different and incompatible ways by different sciences. To make progress in this area, it is important that these key concepts are not just understood at a conceptual level but that they are based in theory and formalized in a consistent way. An example of how to do this well is the work presented to the panel on assessing the value of information in information graphics understanding.

It is important to recognize that HII needs to draw on a vast array of social sciences—including sociology, anthropology, organization theory, psychology, cognitive science, economics, political science, and communication theory. It is unreasonable to expect any one scientist to be aware of fundamentals in all of the social sciences, and collaborators from those fields are necessary. It is important to create a research environment where HII investigators have sufficient access to expertise in the relevant and related disciplines.

The third offset, and HII's relevance to that, provides a tremendous opportunity and the framework in which HII could be particularly valuable to the Army. The breadth of activities is exciting but will require careful management. HII is at risk of having a small number of researchers superficially involved in many areas of research involving humans but not going in depth sufficiently to have a strong scientific impact. An example is autonomous systems—specifically, where could the locus of decision making in an immersive environment reside? To date, there is limited work in this area, and a question is whether greater attention is required in this problem area. Another example is determining which aspects of Army activities are significant—Could HII cover the complete suite of Army activities? Or could the focus be constrained to combat? Developing a challenge problem to focus research for 3 to 5 years would actually enable higher productivity and help focus this area. To be successful in this regard will require leveraging the larger scientific community beyond current collaborative technology alliance (CTA) collaborators and researchers brought in through ARL West.

To accelerate progress, HII needs a showcase problem. Currently, HII is the collaborator on a variety of topics, watches very few of the relevant subareas, and is not regarded as a leader in a given domain. Being the lead on a problem that crosses all three of the identified thrusts is critical for scientific advancement and organizational survival. Possible areas for such showcase problems include developing a theory of mind for HII or developing a theory of cyber-embedded social action. To enable these research efforts will require HII to support a distributed laboratory. A fully informatted distributed space is required—one that is structured to accept information from external information sources, including social media, traditional media, and embedded sensors from different technologies. The ability of the laboratory to interact with this information, with laptops, personal electronic devices, digital agents, and so on, is important for facilitating studies involving groups (not just crowdsourced). Some features of such an environment already exist within the open campus philosophy and the network science research laboratory may serve an important role in this context. The emphasis, however, needs to be on a facility that supports interactions among groups of humans and robots in a truly informatted environment. The HII research program and research team would benefit by aligning with a showcase conference. Regular participation at such a conference will build links with a broader community of researchers to collaborate on problems of interest to the Army. One possible conference is SBP-BRiMS,[2] which already has some military participation, is in the information science area, and draws participants from computer science, engineering, and social science.

ARL needs to focus on developing the right human capital for the HII program. Currently, the HII team is dominated by individuals with computer science backgrounds. Counting both collaborators and members of ARL, there appear to be eight computer scientists, three electrical engineers, three psychologists, and two general social scientists. There needs to be an emphasis in training or recruiting personnel with expertise in sociology, cognitive science, anthropology, or social psychology. Without some in-house expertise, it is challenging to evaluate potential collaborators or watch developments in relevant areas. It is also important to keep abreast of new developments in social sciences—there are a growing number of quantitative, mathematical and computational sociologists, anthropologists, linguists, and cognitive scientists that could bring the needed expertise for HII to be truly impactful. There is insufficient access to that the social science community through the current CTAs or ARL West. In the same context, it may be helpful to facilitate more training for the existing computer scientists and engineers in the relevant social science subareas through short courses and executive education programs. Early career professionals working on transdisciplinary projects also need to receive mentoring from the appropriate branches of computer science and social science. The use of mentors from outside of ARL is also a possibility. While there is evidence of support for training in new computational and mathematical skills, there is less focus on training opportunities in the social sciences. There are mechanisms in place to support collaboration, and these need to be actively pursued to support HII activities. An example of this need is in information visualization, a field where in-house expertise exists at ARL but does not appear to be appropriately engaged in the HII effort. There is evidence that the leadership supports collaboration, and additional attention is required to develop a platform that helps identify partners for collaborative endeavors.

For HII's work to make an impact on the larger scientific community and to be of greater value to Army needs, it needs to have strong theoretical underpinnings that integrate computational and social theoretical constructs. An example of such an approach is the thrust on social computing. The current nascent work in this area is focused on social computing from a largely computer science perspective. This is unlikely to serve HII well because it would retain the existing computer science and engineering emphasis. A transdisciplinary approach that adds relevant and diverse social perspectives would produce the desired integrated theory—which is of greater value to the Army.

This is an exciting new area of research with huge opportunities and strong relevance to the Army. Researchers are enthusiastic about working in this transdisciplinary area but need time to get new training in the human and social sciences arena. The HII research field is vast, and to build prominence, it will be important for ARL to focus and to leverage the expertise in the larger community.

---

[2] SBP-BRiMS is an International Conference on Social Computing, Behavioral-Cultural Modeling, and Prediction and Behavior Representation in Modeling and Simulation.

CYBERSECURITY

Accomplishments

The research projects chosen by ARL seem to be well within the capabilities of the laboratory. Where projects have need for different or more resources than provided by ARL, researchers have shown considerable initiative in teaming with other research institutions. As noted earlier, continuing access to real-world operators and operational data is the greatest resource ARL has in cybersecurity research. However, given their unique capabilities, there are other important cybersecurity areas that ARL is uniquely positioned to address. These are discussed in the challenges and opportunities section below.

The ARL cybersecurity researchers were scientifically competent and in some cases impressively so. Given the scarcity of cybersecurity professionals, this is noteworthy. There is good evidence of collaborative work with other organizations, particularly in the collaborative research projects detailed in the poster session. More importantly, ARL researchers were generally aware of other research in their areas of work. Given the wide range of organizations conducting cybersecurity research—government laboratories, University Affiliated Research Centers, academia, and commercial organizations—this is a difficult, but necessary task.

Stylometry Authorship Attribution for Source Code and Binaries

In cybersecurity, attribution of perpetrators and weapons is a difficult and elusive goal. There have been numerous research efforts to attribute a cyber weapon to a family of cyber weapons of common origin. Most of these efforts have been aimed at attributing source code to common origin. What makes ARL's research novel is being able to carry the work over to binaries. Although the results are preliminary, ARL's research shows promise and is unique among current approaches.

Interrogator Intrusion Detection System

Interrogator represents the best in leveraging ARL's unique dual role in both cybersecurity research and operations. Interrogator is a deployed, widely used tool that has continued to evolve based on feedback from operational experience. Continuing work on Interrogator is encouraged for its contribution to operational excellence, but also for the threads of other research opportunities it uncovers.

Resource Conserving Signatures

The use of Bloom filters as a mechanism for compressing a large corpus of signatures in a network-constrained environment is ingenious. Preliminary testing of the approach with known data sets has shown an order of magnitude bandwidth savings in mobile tactical networks with false positive rates of less than 1 percent. This project is an excellent example of using ARL research to solve a problem that is particularly present in the Army warfighter environment. The approach is novel and the results encouraging. Several posters presented in areas categorized as threat defeat and resilience were of high quality.

Threat Defeat

The overall quality of research in areas related to threat defeat was very good, particularly in malware detection and containment. Even though caution needs to be exercised to avoid over reliance on automated tools in the threat environment in which the Army operates, the ongoing research done in automated detection is considered to be particularly noteworthy.

Understanding the Cyber Threat

The work related to the use of extremely short rules to detect malicious activity and the aforementioned stylometry research to attribute malicious binaries was considered to be excellent and at the leading edge. There were research projects in this category that needed better definition, such as the work related to personality traits and behavior analysis research into aspects of cyber actors and victims. Also, as previously noted, several research projects in this group exemplified the best of the marriage of research and operational environments.

Automated Detection of Hostile Activities

Research related to automated detection is important and impactful. The work related to using Bloom filters to reduce signatures in a constrained network environment is exceptional. Also of particular importance was the research project in cyber-physical security. The security of cyber-physical systems is of immense importance to the Army and the country at large—especially considering the proliferation of devices within the Internet of Things. ARL needs to consider broadening its research efforts in this area, and the ongoing work appears to be a good start. The work in machine learning applications was well organized, but there is skepticism as to how useful the results will be. As an example, the work on grammatical and machine learning used machine learning techniques to address a computationally hard problem. Higher-than-ideal false positives were observed in the approach. Similarly, the work in machine learning for intrusion detection in mobile tactical networks was found to have limited applicability, constrained by the quality of the training set. Conversely, the result from ontological and cognitive modeling was promising and represents a path forward for extension to different threats and network features.

Prevention and Defeat of Hostile Activities

In contrast to the applied nature of research of the previous subsection, the work on preventing and defeating hostile activities was of a slightly more fundamental nature. The work was of good quality and well scoped. It included projects related to evaluation of platform migration defense as a strategy to enhance security of cyber systems. Using simulations, researchers were able to show the influence of platform migration rate and platform diversity on success in defeating attacks. Another project looked at reducing network security problems to graph problems that can be addressed through efficient O (poly) algorithms. Many of the results reported are from early stages of investigation but are nevertheless encouraging.

Resilience

The research projects in resilience appear to have a similar focus to work done at other institutions, but with an Army network constraint. It would have been helpful for the panel to see a typical Army network architecture to provide the context in which this research is applicable. Some of the efforts involve developing tools such as decision support systems that are currently deployed, and use operational feedback to help guide additional research. For example, one decision support system is a red/blue team organization that is called upon to assess the cybersecurity resilience of an organization and provide training to new members. The consolidated virtual inspection program is a methodology to ensure that an existing cyber system is up to date with respect to revision levels of installed systems and existence of appropriate systems. The work facilitates inspection without the physical presence of inspectors and concomitant savings in labor, and has shown promise for widespread implementation. Even though these research projects are looking at deployed technology, it is obvious that applied research to drive the continuous evolution of these capabilities is a worthy goal.

Perhaps the most innovative effort in the resilience area is information security continuous monitoring. Cybersecurity needs ongoing monitoring and continual database updates to meet its operational requirements. Consequently, the database created would be able to scale over time as well as incorporate new metadata. This is required to stay one step ahead of adversaries. Information security continuous monitoring has most of these characteristics. However, the database is currently over the 2-petabyte range and needs to scale to exabytes and beyond. This would require research into scalability of the software and the mathematical algorithms used, as well as the architecture of the underlying hardware platform.

Challenges and Opportunities

For cybersecurity, it is important for ARL researchers to clearly articulate what is different in their environment as compared to a commercial environment. While there is considerable overlap, the differences are important and include issues such as bandwidth constraint, hostile operating environments, and ad hoc joining and departing of networked equipment. These differences could define ARL's cybersecurity thrust. With the emerging Internet of Things, commercial networks are becoming more like the Army's networks.

As noted above, research into the security of cyber-physical systems could be a priority of ARL. Many of the cyber-physical systems in the Army are unique to the Army, and no one else is likely to do research into their security—except perhaps an adversary. The security of these systems is of critical importance to the Army and the nation at large. More research focus is needed on defense against insider attacks. In this context, it is important to recognize how Army insider threats differ from those typically encountered in other organizations. The Army scenario also includes tactical overrun situations, replay attacks, and inside network malware detection. Again, this represents a threat almost unique to Army needs.

ATMOSPHERIC SCIENCES

Accomplishments

The research projects being conducted focus on challenges related to the collection and processing of environmental data from nontraditional observing platforms, the imaging and sensing of aerosols and objects in the battlespace environment, understanding complex atmospheric flows through a combination of field observation experiments and model development, and application of the observations into enabling technologies to enhance renewable energy use. Several of the projects also seek to adapt technology developed for other applications into the domain of atmospheric sciences. Observational field experiments and the proposed deployment of a meteorological sensor array (MSA) at the White Sands Missile Range (WSMR) complex are ambitious projects that seek to push the state of the art in characterizing atmospheric boundary layer (ABL) flows in different types of complex terrain and to eventually model them realistically and accurately. The overall scientific quality of the work is good, and comparable to research conducted at successful university, government, and industry laboratories. Researchers were certainly familiar with the underlying science and cognizant of research done elsewhere. In most cases, the researchers were aware of the potential challenges associated with their projects.

Long-Range Adaptive Passive Imaging System—Development and Field Results

Research on the long-range adaptive passive imaging system is focused on developing approaches to enable enhanced imaging. Specifically, the work seeks to account for the effects of turbulence on images so that objects may be positively identified at significant distances. The approach included experiments to validate and verify theoretical models that have been deployed in a prototype system. The proposed techniques appear to yield a doubling of the range capability, even under degraded conditions. This high-impact result for Army operations has an exceptionally high opportunity to improve in-theater tactical decisions. Collaborations exist with both national and international (e.g., Canadian) defense-related agencies. Even though customer-initiated with a clear application focus, the project entails significant basic science.

Crowdsourcing of Meteorological Data

One of the major challenges facing the Army of 2040 is the characterization of battlespace environments in remote, data-sparse areas where conventional observations are lacking. This research project ties directly into intelligence preparation of the battlespace and aims to leverage future sensor technologies that will be directly embedded within combat systems and on soldiers themselves. Such sensors can automatically collect and transmit real-time environmental data via command and control networks to servers located at a forward operating location that does not have the ability for "reach back" to conventional weather providers, either in rear echelons or in the Continental United States. The project is focused on the back end of the collection process (i.e., server) but utilizes technologies and crowdsourcing approaches that have previously been developed (front-end technologies) in order to obtain data inputs. The server-related research is in its early stages, but results have already shown the high value of the techniques in overcoming data-handling vulnerabilities. In addition, the front-end processes have resulted in one patent and two patent applications. This project is linked to distributed weather-decision support modeling, which concentrates on utilizing multiple sources of environmental information to produce automated guidance on optimal use of Army weapon systems on the battlefield.

Raman Spectroscopy of Atmospheric Aerosols

Given the complexity of the environment and the adequacy of current sensing technologies, there are considerable challenges in detecting chemical and biological agents on the battlefield. A dirty atmosphere can cause signal attenuation for certain weapon-tracking methods (e.g., lasers). The research is focused on developing a Raman spectroscopy-based approach to enable the characterization of supermicron-sized particles and uses an aerosol Raman Spectrometer (resource-effective bioidentification system [REBS]) developed at Battelle. Code was developed so that the REBS could be used as the sampling system for an aerosol Raman spectrometer. Proof of concept has been demonstrated through the identification of bio-like substances, organic substances with C-H stretches, and inorganic salts. The analytical tool was also modified to enable the identification of temporal changes in aerosol constituents. Plans exist to field test the Raman spectroscopy system, leveraging observations from the planned MSA.

The aerosol Raman spectroscopy technique is expected to have significant advantages over fluorescence-based systems for the analyses of biologically based aerosols. Specifically, the Raman-based technique enables positive identification of the chemical composition of bioparticles while simultaneously measuring inorganic aerosol particle composition, a process that is not available with fluorescence-based systems. The most established optical technique that is currently used for the analysis of particles is Fourier Transform Infrared (FTIR) spectroscopy. Raman spectroscopy-based technique has the potential to extend the capabilities of the FTIR techniques by enabling the analyses of particles with compositions that either exhibit high absorption in the infrared range and therefore interfere with FTIR analyses, or may allow for the analyses of particle constituents that are infrared inactive. The Raman-based system has the potential to advance near-real-time optical analyses of particle composition. This work may be especially applicable to and useful for the analyses of bioparticles.

A paper on this work has been accepted for publication in a peer-reviewed journal, and four presentations at national conferences or universities have already been made. These dissemination efforts are an excellent indication of the larger impact of the work.

Advances in Optical Trapping of Aerosols

Real-time detection and characterization of atmospheric aerosol particles in the 0.01 to 10 μm range is an important capability for protecting soldier health and maximizing the effectiveness of battlespace aerosol particle-sensing capabilities. This research has produced highly innovative methods of capturing these particles using laser-trapping approaches. These methods allow real-time analyses of individual aerosol particle composition and optical scattering properties using laser-induced fluorescence and scattering measurements.

Three-Dimensional Polarimetric Imaging

This project exploits recent advances in the development and production of long-wavelength infrared cameras with polarimetric capabilities for real-time image analyses of an object's Stokes parameters. The project currently employs commercially available 7 to 11 μm-band cameras. The initial application has demonstrated successful facial recognition capabilities, even in the presence of heavy makeup. Additional potential applications include complex welding inspections and other materials evaluations.

Phase-Ratio Imaging for Anomaly Detection

This research project is based on a collaboration between ARL and Ukrainian planetary remote sensing researchers. The research has revealed that evaluating the phase ratios of objects in two successive visible images with slightly displaced viewing angles can enhance the detection of surface disturbances. The work is in a preliminary stage and has potential applications in the enhanced detection and recognition of buried mines or improvised explosive devices (IEDs).

Mountainous Slope Transport and Diffusion

Mountainous Slope Transport and Diffusion (MASTODON) is a basic research project that entails a microscale atmospheric observational field experiment that represents a follow-on to the mesoscale field experiment program known as the Mountain Terrain Atmospheric Modeling and Observations (MATERHORN).[3] That project, funded by DOD as a 5-year Multidisciplinary University Research Initiative, is described as the most densely instrumented complex-terrain field experiment in history. MASTODON addresses fundamental scientific questions regarding characterization of ABL flows in

[3] H.J.S. Fernando and Collaborators, The MATERHORN: Unraveling the intricacies of mountain weather, *Bulletin of the American Meteorological Society* 96(11):1945-1967, 2015.

complex terrain under slope heating and cooling conditions, which will be used to improve both ABL and mesoscale numerical weather prediction (NWP) models needed for characterizing the battlespace environment in such areas.

One of the ongoing issues in ABL flow characterization is that traditional slope flow conceptual models under slope heating and cooling conditions do not agree. A number of researchers have investigated this and related scientific problems using a combination of laboratory-scale, thermally driven anabatic flow separation simulations, analyses of prior laboratory upslope flow velocity scaling experiments, and MATERHORN field observations. The laboratory and field data acquired support an upslope flow velocity scaling proposed by Hunt et al. (2003).[4] This is a very promising result that will also have implications for ABL modeling.

There is a large international contingent of researchers, some of whom were involved with MATERHORN and will continue their collaboration in the next field experiment, set to begin in Perdigão, Portugal, in 2017. The Portugal site is unique in the configuration of the terrain with two parallel ridges in relatively close proximity. The experiment is being funded by the European Union as a wind-power exploration investigation and by the National Science Foundation as an investigation of microscale complex terrain processes. ARL has been involved in MATERHORN and will play a prominent role in the Portugal experiment. All of these efforts are pointing toward the technical communities' involvement in the MSA system after its standup at the WSMR.

Distributed Weather Decision Support Modeling

This project can be considered as a front-end component of the crowdsourcing project described previously. This part of the project involves the development of a weather input system as well as a weather-based model on a portable device (e.g., a smartphone). This research and tool development is relatively mature, and has already been transitioned to other DOD agencies. The level of collaboration and technology transition is a clear indication of the profound impact that ARL's weather decision support modeling and associated research has already had within DOD. A further indication of the novelty of the work and broader community impact is evident by the patents that have either been applied for or already granted. These patenting activities are impressive and serve as a clear indication of the exceptionally high level of use-inspired research that has been undertaken.

Weather Impacts on Microgrid Renewable Energy Ramping Event Modeling

A combined field experimentation and modeling effort is under way to enable the efficient use of renewable energy by incorporating weather data into the decision-making process. This novel and highly promising effort is specifically focused on a microgrid scale. Fieldwork was already undertaken in conjunction with collaborators to test the impacts of solar utilization and solar flux measurements on power output from photovoltaic systems. These tests were conducted using large solar panels and enabled the development of operational plans for additional tests as well as appropriate correlations of collected solar flux data for power simulation modeling with the incorporation of meteorological data. Well-conceived plans exist to test additional, flexible solar array systems and to correlate environmental parameters with power output. The work is recognized for its potential impact as ARL investigators have been invited to participate in multiple related scientific panels.

---

[4] J.C.R. Hunt, H.J.S. Fernando, and M. Princevac, Unsteady thermally driven flows on gentle slopes, *Journal of the Atmospheric Sciences* 60:2169-2182, 2003.

The work has clear, positive benefits for Army operations, especially in reducing the frequency for and potential security challenges associated with liquid fuel re-supply convoys, which have been vulnerable to enemy attack. Additionally, the work addresses a critically important, evolving area of renewable energy research for the broader U.S. and international community, especially in under-resourced or remote areas of the world.

Microscale Modeling Focused on Dense Urban and Complex Terrain Domains for Future Army Battlefields

The need for environmental characterization at the microscale level (for the Army, this means spatial scales of 1 to 100 m and temporal scales of the order of minutes) exceeds the capabilities of today's mesoscale NWP models. Previous work in this area has resulted in a diagnostic wind flow model known as 3DWF, which has no thermal forcing but can be run on horizontal scales small enough to include the presence of individual buildings. The 3DWF has been transitioned to several operational users but lacks more sophisticated physical processes such as thermally induced slope wind flow.

The focus is on developing a suite of atmospheric boundary layer environment (ABLE) models to predict mean wind, temperature, moisture, and turbulence over urban and complex terrain in near real time. Three following candidate models for ABLE are under consideration: (1) ABLE—Computational Fluid Dynamics, a finite volume model where the emphasis has been on the treatment of lateral boundary conditions in the model; (2) ABLE—Vortex Filament Scheme, which has no thermal forcing but the use of a "thermal bubble" perturbation appears to show some promise; (3) ABLE—Lattice Boltzmann Method (LBM), which employs LBM to retrieve macroscale features from the motions of individual molecules. The ABLE-LBM model can account for different heating perturbations and distributions of stability and can model the flow around complex terrain and buildings. It has undergone preliminary validation by use of elementary benchmark flows to determine if it reproduces the expected flow results.

Use of a New Hybrid Nudging/Variational LAPS Approach for WRF-ARW Initialization

The operational requirements for this project are driven by the need for high-resolution battlespace environmental characterizations in remote areas where conditions are not practical for conventional "reach-back" strategies to obtain weather forecast information from rear-echelon or Continental U.S. locations. These environments can be quite complex and variable—from rugged terrain to urban canyons—and provide the backdrop for the hybrid nudging/variational local analysis and prediction system (LAPS) project. The model chosen for development is a community-developed mesoscale NWP model known as the weather research and forecasting (WRF)—advanced research WRF (ARW). The WRF-ARW system, as modified by ARL, is known as the weather running estimate—nowcast (WRE-N); the system is expected to run over small geographic regions (100 to 500 km on a side) for very short forecast timeframes (1 to 6 hours). An additional requirement is for the WRE-N to be run every 30 minutes, making the assimilation of indirect and asynchronous observations from a variety of sources an issue of highest importance.

The investigators are working with other NWP modeling groups in the National Oceanic and Atmospheric Administration and the university community to address the unique requirements of assimilating such data in the computationally constrained environment of a forward operating base. The computational constraints do not allow the employment of more state-of-the-art data assimilation techniques, so the group is examining combinations of already-proven assimilation methods such as observational/analysis nudging and variational data assimilation. Another unique aspect of this project is the approach that variational LAPS uses, which proceeds from larger to smaller spatial scales and is computationally faster than more advanced methods such as Ensemble Kalman Filtering and 4 Dimensional Data Assimilation.

Several different variations of the assimilation technique, including variational LAPS alone, nudging alone, no assimilation, and combinations of variational LAPS and nudging or the hybrid scheme, were assessed in numerical simulations using existing recorded data. The hybrid scheme showed some promise when compared to the other initialization methods by better capturing several areas of deep moist convection, but the approach is not without its flaws.

Assessment of Storm-Scale Convective Forecasts

One of the most important considerations in battlespace environmental characterization and prediction is determining the quality of predictions. While commonly used statistical forecast verification scores capture important information, they do not provide insights into a model's ability to capture specific weather phenomena accurately. It is well known that coarser-resolution NWP models can actually have better statistical verification scores than finer-scale models due to the "smoothing out" of fine-scale features that the higher-resolution models may capture, but they are deficient in predicting the correct location and/or time frame. The research seeks to address this deficiency by using an object-based verification scheme based on the method for object-based diagnostic evaluation (MODE), developed at the National Center for Atmospheric Research.

The MODE technique was tested against data for a 3-week period of High-Resolution Rapid Refresh (HRRR) forecasts in 2010. The MODE technique verified deep moist convection forecasts from HRRR by comparing simulated vertically integrated liquid (VIL), a commonly used radar-based diagnostic for deep moist convection, to observed VIL available at 15-minute intervals during the verification period. The VIL is a discontinuous field, and a threshold value was used to employ the MODE technique. The results of the verification study revealed some interesting characteristics about WRF's ability to develop and maintain deep moist convection and how these differed by time of day, size of convective system, and geographic region. The plan is to adapt this technique to verify other atmospheric parameters from other model forecasts such as WRE-N. This is noteworthy but not without attendant challenges.

Design of Experiments for Verification and Assessment of Fine-Scale Atmospheric Forecasts

This research project seeks to determine the optimum configuration for running a mesoscale NWP model based on a statistical technique that has been used in other modeling and simulation applications. At present, the choice of numeric and physics options when configuring an NWP model run is subjective and largely based on anecdotal accounts of which schemes work best in certain geographic regions and climatic conditions. The research seeks an approach to quantify this information in order to produce a type of NWP model configuration "playbook" that can be employed in theater by deployed personnel who are not modeling specialists but nonetheless need the information that models such as WRE-N can provide. There appear to be potential synergies between this project and the object-based verification method discussed above, since the model forecast verification is key to proper employment of this statistically based model configuration scheme.

Sub-km Resolution WRF-ARW Modeling Studies During MATERHORN Intensive Observing Period 6

This is a project with connections to MASTODON and the two previous projects. In this investigation, WRF-ARW was run on a 0.5-km horizontal grid over the MATERHORN field experiment region during intensive observing period 6. The results of the model run were mixed. As the model was run without data assimilation (cold start), the complex terrain flow features during the first diurnal heating and cooling cycle were not captured adequately, but features such as the Playa breeze were partially captured during the subsequent model diurnal cycle. However, other types of complex terrain-induced flows, such as slope flows above the surface, were not reproduced accurately in the model simulation. Future plans include the adoption of a large eddy simulation transition approach for running the model at smaller horizontal grid spacing.

Challenges and Opportunities

In the long-range adaptive passive imaging system project, the broader fundamental scientific applicability and path for future project development are somewhat unclear. Indications of how the proposed effort might be integrated into other Information Sciences Campaign efforts would be ideal and may lead to further opportunities to extend the longer-term value of the existing effort.

Several opportunities exist to expand the Information Sciences Campaign's efforts in the crowdsourcing projects. For example, questions of how uncertainty in data, data analyses, and subsequent error propagation would be handled are important in the development of broadly applicable research. Articulating how these issues of uncertainty and quality control would be handled or, alternatively, how fundamental research would progress in these areas is a larger issue that could be more thoroughly addressed in this research effort. Additionally, an ongoing challenge to the crowdsourcing effort is the identification, calibration, and deployment of compact, low-power, and robust environmental sensors that will fill its front-end data stream. There are numerous unknowns that may potentially exist with sensor quality and functionality, thereby resulting in data quality challenges. Early collaborative efforts in the simultaneous development of the sensors and data acquisition and modeling techniques may help to improve compatibility between data acquisition and data handling, as well as enhance quality control. Collaborative engagements are encouraged not only across ARL and other DOD agencies, but also with universities and private industry.

The Raman spectroscopy of atmospheric aerosols project is promising research but not without questions. First, it is unclear how this method compares to existing particle analyses techniques. In order to gain broader scientific acceptance of a Raman spectroscopy-based system, ARL staff are encouraged to engage in a thorough comparative study using other aerosol characterization methods and systems. Future work to expand analyses down to the submicron size range may present an additional opportunity to extend the applicability of the Raman spectroscopy-based analyses of particles. In particular, focusing on nano-sized particles would enable novel research in the area of nanoparticle toxicity. The latter is an important concern for the Army. Additional information on how complex matrices (that would be more representative of an actual battlefield environment) would impact the analyses can further extend the applicability of the project. A unique opportunity exists to further extend the work to characterize the oxidation of organic particles, and specifically the shifting of C-C or C-H bonds to C=O and COH bonds, to enable the identification of oxygenated compounds that are typical of secondary aerosol formation.

The advances in the optical trapping of aerosols project is currently at the successful proof-of-concept level and may very well become a useful laboratory tool for real-time detection and characterization of a whole variety of aerosol particles. Possible evolution into a useful field measurement tool for force protection and situation awareness will require strong electro optical, mechanical, and automated instrument operation and data analysis software advances, as well as instrumentation integration engineering activities.

The utility of the three-dimensional polarimetric imaging technique in the advancement of atmospheric science is not obvious. Over the past decade, NASA has sponsored some work in remote sensing from airborne or satellite-based polarimetric imagers to assess atmospheric aerosol shape parameters, using instruments in the visible and near-infrared spectral bands. While the phase-ratio imaging for anomaly detection technique may be useful in detecting disturbances in battlefield terrain, such as mine or IED burial scars, its utility for atmospheric measurement has yet to be identified.

The MASTODON project is one of ARL's most ambitious—with potentially the highest payoff. It is linked to the installation and operating of the MSA, which has made considerable progress in terms of equipment funding and purchase since the 2013-2014 review.[5] The MSA promises to be a cornerstone for the success of the Atmospheric Sciences Center at WSMR. However, a significant commitment of resources will be needed for continuous operations and maintenance of the sensor array, once installed, in order to attract world-class researchers to the facility to use the data and collaborate with ARL scientists. At the time of the panel's visit, siting approval had not yet been received from the WSMR leadership.

There are significant leadership and collaboration opportunities for ARL by leveraging the project on weather impacts on microgrid renewable energy ramping event modeling. ARL project personnel are encouraged to continue engaging with the technical community, particularly in terms of incorporating the next-generation innovations in renewable energy technologies. This will enable rapid integration of new system parameters into the modeling effort. Innovations in the devices and techniques may, for example, enable more effective use of a wider range of the electromagnetic spectrum, thereby enhancing solar use efficiency and, ultimately, energy availability and power output. As renewable energy technologies change, the weather-renewable energy modeling tools will need to be flexible enough to adapt. Enough information on the modeling component of the project was not available for the panel to effectively assess how variations in renewable energy technologies would be incorporated into the efforts. It was somewhat unclear whether the project team has a plan to enable additional engagement from the renewable energy side of the project, although it was clear that efforts are under way to establish cooperative agreements with at least one university partner on the weather forecasting side. Cooperative agreements with university and industry researchers who are engaged in renewable energy projects are strongly encouraged. There are also numerous international institutions—particularly in Latin America—that have a strong interest in coupling environmental issues and renewable energy on the smaller scales that the ARL project team is undertaking.

While the use of elementary benchmarks for verification in the microscale modeling project is encouraging, this process will become more challenging as the developers move into more complex scenarios. Additionally, there is the question of how to model turbulence, which is of great importance in transport and diffusion of chemical and biological agents as well as smoke and other obscurants on the battlefield. The investigators mentioned the desire to use wind tunnel experiments for model evaluation, an approach that is to be encouraged. Opportunities may also present themselves in the area of code optimization. The LBM technique lends well to parallel processing, which will be important as the modeling system begins to approach readiness for technology transition. This is a relatively new approach, and close collaboration with researchers in the high-performance computing area is encour-

[5] National Research Council, *2013-2014 Assessment of the Army Research Laboratory*, The National Academies Press, Washington, D.C., 2015.

aged. An additional consideration is the use of the MSA data for model verification; the investigators need to take advantage of this unique data source.

The hybrid nudging/variational LAPS project has similar challenges to the microscale modeling project regarding validation and verification. Given the requirement to run WRE-N anywhere in the world, it will be important to have as complete an evaluation as feasible, given that the model cannot be pretested for all locations at all times of the year and in all climatic conditions. There is a great deal of qualitative and quantitative verification required for the hybrid assimilation scheme. For the test case where measurements from the 2013 Moore, Oklahoma, tornado were used for model assessment, the hybrid scheme did produce a better deep, moist convection forecast in comparison to the other methods. In several portions of the model domain, however, it performed poorly; for instance, it did not reproduce the convective system that spawned the Moore tornado itself. This brings up the question of accuracy versus realism. The model results show WRF-ARW has the ability to reproduce fine-scale features of a mesoscale convective system realistically, such as the strong updraft and hail-producing regions. However, for present and future Army needs, the model needs to be able to reproduce such features in the right location and at the right time. Additionally, the assimilation technique needs to be tested in more data-sparse areas and for other output parameters of interest to the Army (e.g., surface temperature, low-level winds). There is a great opportunity to use this modeling system to test the efficacy of future observing systems, such as soldier-borne sensors and various types of unmanned aerial systems, through the use of observing system simulation experiments. The investigators are aware of these challenges and opportunities, and their collaborations with other NWP modeling groups are very encouraging. An additional consideration is the use of the MSA data for model verification, and the investigators are encouraged to take advantage of this unique data source.

The assessment of storm-scale convective forecasts project has potential to add to the atmospheric science community's knowledge of NWP model verification through the unique approach of phenomenon-based quantitative evaluation. The expansion of the project to other models (e.g., WRE-N) and other parameters will be challenging simply due to the nature of moving from one software system to another. A related issue will be the development of an appropriate protocol for choosing the right variables at the right locations, since it is not possible to prerun the model and verify it everywhere at all times and in all climatic conditions.

The design of experiments for the assessment of fine-scale atmospheric forecasts may be the highest-risk, highest-payoff project that was reviewed by the panel in the current cycle. The technique has never been applied to an atmospheric NWP model, and there is a dearth of literature on the subject of choosing numeric and physics packages in an NWP system based on how each influence the overall accuracy of the model predictions. As they attempt the translation of this proven statistical technique to the atmospheric modeling problem, it will be critical for investigators to reach out to both the operational and research components of the NWP community to get as much information as possible. Successful development of this technique has the potential to influence the operational NWP centers around the world; having a robust set of model-configuration guidance (playbook) for deploying forces to use will be a separate, but equally daunting, challenge.

The subkilometer-resolution WRF-ARW modeling studies project had some encouraging results, but like the hybrid nudging/variational LAPS project, more needs to be done in terms of model verification and potential enhancements that could improve the model forecasts. As stated for the latter project and the microscale modeling project, the use of MSA data for model verification will be crucial in advancing the goals of this research. It is important to get a sense of the ability of the model to reproduce the small-scale flow features that are critical to battlespace environmental characterization and prediction.

An overarching positive observation about the work in this area relates to the amount and extent of collaboration taking place within the laboratory (through the science and technology or S&T campaigns) and (very importantly) with knowledgeable research groups both in the United States and internationally. These collaborative efforts were highlighted numerous times by the project investigators. As an example, in the five modeling-related projects reviewed, collaborations involved federal laboratories and several universities. The observational field experiment projects similarly had notable collaborations with national and international universities.

OVERALL QUALITY OF THE WORK

The research portfolio of the Information Sciences Campaign reviewed in this current cycle was expansive and covered diverse areas. The projects reviewed range from those advancing fundamental science to those focused on enabling technologies and applications. The ongoing projects demonstrated relevance to the future missions of the Army and were generally of good technical quality. There are additional opportunities to further drive scientific innovation through enhanced integration and collaboration across campaigns.

ARL has focused on increasing the number of Ph.D. scientists on the research staff in critical areas of expertise, and this has had a measurable impact on the overall quality of technical work. In the 2013-2014 review, it was noted that efforts in the network science area would be enhanced by bringing greater technical diversity to the workforce—in particular, added strengths in the social and mathematical sciences. This diversity and added strength in the social sciences is also critical to the HII initiative. Among the ARL researchers, there was a good awareness of external research and connections to professional organizations and external research communities; research results are appearing in respected conference proceedings and in archival journals. There is room for even broader dissemination of these results to a larger scientific community. As noted in earlier reviews, the mission-oriented thrust helps differentiate the ongoing research from efforts pursued elsewhere and creates opportunities for impactful technical contributions. The impact of the work can be further enhanced by clear articulation of unique, cutting-edge research questions.

Given the vast range of computing available today, especially for data-intensive work, it is not clear whether ARL has the resources needed to carry on the current work as projects move past the proof-of-principle stage. It would be good to consider whether the resources available, or planned to be available, will be adequate for the next 2-3 years. It will be important to assure with confidence that the following questions are being asked and adequately addressed: Is there a systematic approach to estimating the need? Is there a need for occasional access to very large-scale parallel environments (such as a 10,000 or more node system—that is, a 250,000+ core system)? Does ARL need to develop partnerships with providers of very large scale, data-intensive systems, either with other federal agencies (e.g., NSF or DOE) or with private industry (e.g., Amazon or Google)?

Computing is going through a transition (including post-Moore's law and the looming end of CMOS scaling, a revolution in data-intensive science, and the promise of quantum computing, which may only be useful for some niche areas but those include ones of great importance to ARL). This transition makes it difficult to make firm long-term plans, but it is essential to develop a strategy to address the possible directions in which computing may go. ARL needs to perform some research to better understand the consequences of these changes for 10 and 20 year strategic plans.

The research portfolio in SIIS includes a growing component focused on enabling technologies, the underlying science, and novel applications to intelligent and autonomous systems. In all cases the intelligent systems issues under study showed clear relevance to the future missions of the Army. The projects were generally of good technical quality. Importantly, especially among the junior researchers, there was a good awareness of external research and connections to professional organizations and outside research communities; these are important for maintaining and growing the technical quality of the research. The research results are appearing in respected conference proceedings and archival journals. ARL has continued to demonstrate its responsiveness to the general comments and recommendations of the 2013-2014 review and is increasing the number of Ph.D. scientists on its research staff in key areas.

The opportunity for strong technical contributions and for differentiating the work from research conducted elsewhere, as well as enhancing the value proposition for the Army, lies in a mission-oriented thrust to the research. A number of projects in the research portfolio have just such a mission focus and the associated constraints (e.g., limits on the volume of prior information or on available network bandwidth), and these serve as a clear driver of the technical direction of the work. Other projects would similarly benefit by a focus on those areas where the special needs of the Army are not addressed in the basic research agenda being pursued outside of ARL, including the development of technology products related to this research.

The research thrust in sensing and effecting includes projects that address emergent needs of the Army. New theoretical advances and resulting tools have led to a rapid evolution of technology in this domain, and the 3- to 7-year horizons for ongoing research projects at ARL are quite reasonable. The research was generally of high quality, with a focus that is unlikely to be pursued by researchers at universities or at other federal or industrial research laboratories. As an example, recognition of ease of disruption and cost of errors (e.g., human life) needs special attention; field data become critical to the solution of this problem. The researchers generally demonstrated a good understanding of the problems being considered, were able to provide an appropriate statement of the research problem, and pursued appropriate methodologies. They demonstrated awareness of the state of the art and of the related research pursued elsewhere. The facilities required to support the research, including both instrumentation and the computational tools, were adequate.

Research projects in the area of networks and communications were generally of high quality and relevant to the needs of the Army. The research results are being published in archival journals and in respected conference proceedings. The recent relaxation in rules permitting travel to relevant scientific conferences will have a continued positive impact on the overall research environment. The research portfolio in NC has a focus on both theoretical research and on applications and technology demonstrations. The portfolio would have enhanced impact if, in addition to narrow and focused projects that seek answers to very-well-defined scientific or technical questions, it also included projects that are broader in scope that address a specific Army need. In general, researchers in the portfolio demonstrated good awareness of related work in other research communities. There was, however, in some of the interdisciplinary research projects, uneven awareness of the literature and of accepted approaches in constituent disciplines. Additional focus on formulating relevant research problem statements would be beneficial for these interdisciplinary research endeavors.

The research focus in the area of HII is very recent and includes projects that largely involve the work of collaborators. Existing projects in related areas have been transitioned and refocused on HII only recently. It is, therefore, too early to provide a definitive assessment of the overall scientific quality of the research. The work reviewed was comparable in effort, scope, and creativity to much of the work in universities and government and industry laboratories. However, with the exception of the projects identified above, the research that was presented was not well connected to much of the existing work that is being pursued elsewhere. Many researchers were unaware of the human-centered challenges, the relevant social, psychological, and cognitive theories, and the associated methods. They had difficulty articulating specific human-centered research questions and associated challenges. Many of the researchers were unaware of the relevant theories and methods, and tended to make assumptions about human behavior based on personal experience rather than sound and verified theories. In many cases, the posters presented to the panel did not do justice to the research. They often did not provide information on the fundamental research question, the pertinent state of the art, whom the research would impact, and the key results. In other words, the posters rarely answered the Heilmeier questions.

The overall research program in cybersecurity was well aligned with the current and future needs of the Army. Notable progress has been made since the last review. The overall quality of research was mixed with some of the projects rated as truly excellent. The goals, plans, and approaches of the various projects were notably well defined, and the methodologies employed were appropriate for the research undertaken. The research projects appeared to be well conceived with well-defined goals and realistic projects plans. Both the posters and the presentations were generally clear and understandable. In particular, the projects seemed to be adequately scoped to the resources and available time. As expected of such a diverse collection of research topics, there were a wide variety of methodologies and scientific approaches evident in the project presentations. The majority of the projects showed suitable scientific rigor and practice. However, the methodologies in some of the research utilizing psychosocial aspects of adversary actions seemed ill-defined and unconvincing.

In work related to atmospheric sciences, the overall scientific quality of the research was good and comparable to research conducted at successful university, government, and industry laboratories. Good progress that has been made since the 2014 review, with multiple projects moving in the right direction. Researchers were familiar with the underlying science and cognizant of research done elsewhere. In most cases, the researchers were aware of the potential challenges associated with their projects and had given considerable thought regarding ways to address questions in that regard. For example, in the hybrid nudging/variational LAPS project, the performance of both data assimilation techniques is dependent on which problem parameters are chosen, such as the weighting coefficient in the nudging method. The choice of such methods is an art, and the researchers recognized it as such. In the microscale modeling project, the researchers understood the limitations of the various approaches that they are investigating and are pursuing creative and innovative solution strategies to solve a complex flow problem. Furthermore, researchers are exploring methods, such as the vortex filament approach and the Lattice Boltzmann Method, to capture complex physical flow phenomena and attain greater computational efficiency. The amount and extent of collaboration taking place within the laboratory (through the S&T campaigns) and with knowledgeable domestic and international research groups is truly commendable. Several of the collaborative efforts are a direct result of the open campus initiative, which has brought postdoctoral researchers into the laboratory, several of whom have been hired as term employees and are continuing their work and their research relationships with their respective alma maters.

5

Computational Sciences

The Panel on Information Science at the Army Research Laboratory (ARL) conducted its review of ARL's advanced computing architectures, computing sciences, data-intensive sciences, and predictive simulation sciences at Aberdeen, Maryland, on June 15-16, 2015. This chapter evaluates that work, recognizing that it represents only a portion of ARL's Computational Sciences Campaign.

ADVANCED COMPUTING ARCHITECTURES

Alongside a focused research effort in tactical high-performance computing (HPC), researchers have moved to explore new and interesting computer architectures, including neuro-synaptic computing, the epiphany many-core chip, and quantum networking. The research staff in these areas of work exhibited high quality research skills and is well organized and positioned to succeed.

Within computational sciences, big data/analytics and HPC are treated separately. Given that important applications will increasingly combine high-performance analytics and high-performance computing, it is important to assess the impact of integrated computing on architecture and operating environments. Architectures will move to datacentric designs where compute is moved to data throughout the systems environment. This will be true in the data center and at the tactical edge. Analytics on very large dynamically changing graphs will be critical to gaining insights from big data. While the computation in scientific computing can be statically scheduled, computing on graphs is data-dependent. The operating environment therefore needs to change to schedule dynamic computations; this capability is also required for resilience in the battlefield.

Accomplishments

Work in tactical HPC has identified key aspects of the solution space and is exploring important issues related to establishing the utility of those aspects. A particularly interesting observation by ARL researchers is that the tactical HPC space is potentially richer than the commercial space, owing to differences such as communication reliability and potential deployment of field resources. The commercial space currently consists of handheld devices and cloud-computing resources deployed in data centers. The tactical HPC space includes handheld devices, field-deployed tactical (potentially mobile) resources, strategic resources close to the battlefield environment, and cloud-computing resources located in military data centers. ARL has developed models that will guide offloading decisions in these complex computing environments. An interesting question for ARL to explore is the commercial applicability of the intermediate resources to issues such as smart highways.

An example of tactical HPC is the project in multiobjective geometric optimization, a tool to safely guide soldiers to strategic viewing points in the presence of adversaries at known locations. The computational effort (ray tracing) to determine, in real time, a safe path in a 3D modeled megacity, avoiding detection by one or more known enemies, is significant and beyond the capabilities of current and emerging handheld devices. Input from the handheld device (tablet) to the remote HPC system and receipt of output back to the tablet do not place excessive demands on the energy-limited tablet. Researchers have explored a variety of solutions on a set of HPC platforms, with varying performance. This is a high-priority Army application, but security and resilience have yet to be considered.

The work on computation ferrying is exploring an important aspect of realizing ARL's vision of tactical HPC. It considers the tasks that could be computed on handheld devices or on mobile tactical HPC resources. The project has developed a simulation environment in which to study the behavioral properties of the system to determine the specific scenario parameters for when it is necessary and reasonable to use offload tasks from the handheld devices. Initial results have demonstrated that offloading proves most useful for intermediate task sizes, which achieve a balance between communication cost and improved computation speed. Overall, this direction will motivate an end-to-end solution for battlefield computing.

In addition to establishing modeling and simulation capabilities to guide offload decisions, ARL has explored critical issues related to programmability and performance of edge (handheld) devices. The work with open computing language (OpenCL) programs demonstrates that devices can achieve significant computational efficiency, particularly in terms of power and performance. This work can serve as the basis for parameterizing the models and simulations that will guide offload decisions. A possible direction for future work on handheld devices is to develop an understanding of the trade-off between performance per watt (i.e., power efficiency) and the amount of power available in handheld devices (i.e., battery lifetime of those devices).

The project in dynamic binary translation (DBT), which demonstrated the fastest cycle-accurate emulator design in the 2014 MEMOCODE[1] contest, falls in the areas of power-performance and heterogeneous computing. The research group developed a DBT to allow fast cross-architecture execution of binary codes. This would, for example, allow legacy binary code developed on a now obsolete architecture, e.g., i8080, for which the source code is not available, to be run efficiently on another current architecture, e.g., a low-power ARM.[2] DBT is a well-mined area with results in the literature dating

[1] MEMOCODE is the Association for Computing Machinery (ACM)-Institute of Electrical and Electronics Engineers (IEEE) International Conference on Formal Methods and Models for System Design.

[2] ARM is a family of instruction set architectures for computer processors developed by British company ARM Holdings, based on a reduced instruction set computing (RISC) architecture.

back to the 1990s. What sets this work apart is that this DBT outperformed all other MEMOCODE entries by a factor of seven, with close to 1.1 cycle efficiency.

Challenges and Opportunities

In advanced computing architectures, the Computational Sciences Campaign specifically identified the goal of leadership in areas that include heterogeneous computing; many-core integrated architectures; tactical HPC; power, performance, and portability; quantum computing; neurosynaptic computing; and software-defined networking (SDN). Leadership is a posture in which ARL maintains considerable in-house expertise, has a substantive infrastructure, and devotes significant investment based on unique Army needs. If one adds to this objective a leadership role in the research community, then the above list is extremely ambitious. It would be sensible for ARL to define a more limited list of areas in which it will achieve this leadership.

ARL is developing in-house expertise with potential to establish a worldwide leadership role in neurosynaptic computing and has plans to establish substantive infrastructure by investment in the area. Additionally, it is investing in quantum networking as a specific focus within the topic of quantum computing.

There is significant research effort worldwide in heterogeneous computing, many-core integrated architectures, and power, performance, and portability. Most of this work is taking place within data centers, with some researchers exploring deployment in the field. ARL has a clear opportunity for broader community leadership and significant impact on the Army's capability if research in these areas is focused on tactical HPC. As part of this work, ARL could investigate advanced techniques for security and resilience in the battlefield infrastructure. The battlefield environment implies much higher costs for usability and reliability than does a commercial environment. The investigation here could look at the rescheduling of work dynamically as parts of the infrastructure go offline and come back online.

The majority of the posters presented as part of advanced computing architectures can be brought under this umbrella. While the posters mostly represented good work, many were missing the driving application or scenario, and they tended to be incomplete. The two posters related to SDN (software-defined networking) were missing a critical piece—the dynamic reconfiguration of the network has to be driven by the requirements of the workload, which need to be specified in declarative language or detected automatically. The ARL researchers are to be commended for their participation in the GENI (global environment for network innovations) community and for the quality of the infrastructure that they have built.

The two key challenges for tactical HPC are (1) the selection of high-priority Army applications and scenarios to drive the research and (2) the inclusion of security and resilience of the infrastructure as an explicit research goal. The latter challenge may necessitate new hiring and/or the forging of external partnerships.

Broader community leadership in quantum networking is also possible, but a more in-depth review of this thrust needs to be considered. The research is being done in the context of a larger Center for Distributed Quantum Information that has the specific objective to study the essential elements for implementing and exploiting a resilient network of quantum devices. The level of resources allocated needs to be reassessed if the goal is to build leadership in this area.

ARL researchers are pursuing interesting and important research in the area of advanced computing architectures. However, the work frequently is not appearing in tier one conferences and publications—e.g., International Symposium on Computer Architecture; IEEE's Micro's Programming Language Design and Implementation (PLDI), Architectural Support for Programming Languages and

Operating Systems (ASPLOS), and SC (formerly Supercomputing); and the International Conference for High Performance Computing, Networking, Storage and Analysis, which are the most important publication venues for research in the area. The key issue for achieving such publication successes for ARL is learning how to package research results for those venues. ARL researchers can only learn this skill by attending those conferences and participating in their organization, such as program committees.

COMPUTING SCIENCES

Computing sciences aims to develop the understanding, tools, techniques, and methodologies to fully exploit emerging computing architectures. This is accomplished through the realization of efficient task parallel algorithms and the use of advanced memory hierarchies. These efforts are expected to greatly reduce the time required to restate algorithms in parallel form and to correct implementation faults and bugs. The research areas of focus can be broadly categorized as programming languages, computational environments, and software integration. More specifically, research in programming languages seeks to improve performance, portability, and productivity of methods for Army-specific applications. Software integration aims to reduce the cost of entry for developers of scientific software and for scientists, engineers, and other users, allowing them to effectively use evolving computing environments. The evolution of the computing infrastructure is creating new challenges, ranging from energy-aware software development to software for massively parallel and distributed systems. The research in software integration is focused on addressing these challenges and developing an approach for software design that is predicated on the principle of systematic software reuse.

Accomplishments

The ARL computing sciences group has established a strategic focus in quantum computing; parallel processing environments for large, heterogeneous parallel systems; and tools to simplify application development for HPC environments. Establishing a coherent and sustaining strategy in these areas was a major accomplishment. The group has also grown its work in exploratory research while still providing support for issues of immediate importance to the Army's mission. The group has established an approach to keeping its computing infrastructure up to date with the latest hardware and software systems and tools, effectively upgrading the HPC system infrastructure every 2 years (staggered replacement of two HPC systems, each on a 4-year replacement cycle). While this turnover in equipment can be a challenge for application developers, the group is working to provide tools to simplify application development and help the developer optimize performance on the more advanced systems.

Research in quantum computing and software environment optimization is especially noteworthy and recognized as being leadership work that advances the basic science in important areas of computing technology. The focus within quantum computing is hardware abstraction at the function level (referring to a network of devices rather than an individual device) versus the Qbit, or device, level. This work has the potential to advance the knowledge and technology for future quantum computing systems.

Similarly, the focus on software environment optimization for hierarchical many-core and heterogeneous processing environments is on tools and algorithms to support code development and optimization of emerging HPC system structures. This work will be important for the efficient and effective use of HPC systems under development for ARL applications.

Two projects stood out as incorporating outside-the-box concepts and resulting in high-quality basic research that integrated theory, computation, and experimentation. These projects were focused on the development of threaded message-passing interface (MPI) for reduced instruction set computing (RISC)-

array multicore processors and on HPC scaled quantum hardware description language (QHDL)-based modeling of entanglement dynamics in Jaynes-Cummings circuits in open system evolution. The first of these provides ARL a solid foundation to pursue innovative solutions to the challenging problem of power-efficient parallel programming. The second is one of the few efforts in quantum networking, and the development of hardware abstraction language will have strong value in future systems. Another project, dealing with an auto-tuning benchmark for HPC accelerators, represents valuable tactical research versus long-range basic science. This research supports initial performance optimization of existing applications and benchmark codes on emerging HPC systems. Auto-tuning in this work is a higher level tuning (as contrasted with a widely practiced industry approach of using libraries).

Work related to communication-avoiding approaches for Lanczoz eigensolvers demonstrated a good understanding of the interaction of mathematics and HPC. The approach is in the process of being submitted to Sandia National Laboratories for inclusion in production codes (Anasazi) to realize the benefits of scaling; the potential for such deployment is a sign of high quality and immediate scientific relevance and impact. A project dealing with hierarchical multiscale computational framework was technically strong but not considered as representing basic research. This work could be transitioned to a production code.

Challenges and Opportunities

The major challenges facing the computing sciences group include scaling applications across large parallel heterogeneous systems, developing tools and frameworks that work on multiple parallel system architectures, and improving software performance and developer productivity in the face of all evolving system architectures and capabilities. System designers have to create more exotic system architectures to increase system capability and performance, since processor clock speed has been flat for many years. Also, new and emerging applications target the processing of very large data sets or cover a large problem space (e.g., time and structure size), putting pressure on system designers to consider system architectures optimized for datacentric computing (large data/computation ratio). The complexity of system architectures and the need to handle large data sets have significantly increased the challenge of giving application developers effective and efficient software environments, frameworks, and languages. However, ARL's computing sciences group has the opportunity to create the tools needed to optimize performance and scale for mission-critical applications and user environments. Increasing the capability of the Army is dependent on the computing sciences group's development of technologies to continue scaling of capacity and performance of the Army's computing infrastructure and emerging applications.

Research related to domain-specific languages is an appropriate focus given the promise of improvements in productivity. Work is ongoing to create a new language (e.g., Liszt) for domain-specific parallel programming. The analysis of results did not include comparison with other graphics processing unit (GPU) languages, and the metric used to evaluate productivity (e.g., lines of code) was not sufficient. It was not clear if this work holds promise of transitioning to an ARL customer or to any outside agencies, given the lack of control of GPU architectures. It was also unclear whether work on datacentric, extreme-scale computing provides a viable, sustainable, or unique capability over existing task-oriented parallelism frameworks. Related work on adaptive fast multipole methods using task-oriented parallelism suffered from a lack of supporting performance data and insufficient analysis and articulation of scaling properties. Work related to the implementation of a Bayesian quantum game builds an experimental, computational, and theoretical framework for using quantum statistics to model aspects of cognition. While the approach is conceptually promising, a more detailed presentation and references to related work would be beneficial.

DATA-INTENSIVE SCIENCES

The goal of the data-intensive sciences research portfolio is to understand and exploit the fundamental aspects of large-scale, multidimensional data analytics. There are currently three areas of focus in data-intensive sciences research: the science of large data, computational mathematics for data analytics, and real-time data access and analytics. The first two of these research areas were part of the current review.

Accomplishments

Current accomplishments in this research portfolio include new model order reduction methods for partial differential equations (PDEs), cognitively steered exploration of solutions to PDEs at different resolution, efficient summarization and visualization of high-dimensional data sets, and concise characterization of spall damage in materials. ARL has a goal of leading research in the areas of explosive mechanisms and high strain rate and fracture. A new research effort has been prepared in computation of game strategies exploiting symbolic representation and in using a new neuromorphic computational architecture.

It is encouraging to see the focus in these important areas. The initial set of projects appears to be promising and is beginning to show good progress. Specifically, the work on developing a high-performance, sparse, nonnegative, least square solver advances the state of the art and leverages ARL's expertise in numerical analysis and HPC. Similarly, the work on neuromorphic computing represents a fresh and original approach.

The work related to model order reduction methods for large-scale simulation data is part of a larger collaboration with the Stanford University group on model order reduction. ARL is pioneering the parallelization of hyperreduction methods to bring greater computational efficiency in the simulation of PDE codes. The approach being implemented is novel and important. The researchers are well informed on the state-of-the-art approaches in this field, and appropriate equipment and numerical models have been selected. The availability of world-class HPC resources is a strong driver, and the collaboration with the Stanford University group is impressive. The work contains the appropriate mix of theory and computation. Additional work in deriving error estimates and bounds for the hyperreduction results would strengthen the impact of this work.

The new thrust in programming approaches for neuromorphic cognitive computing represents a novel and intriguing research program. The principal investigator has the appropriate background for this work and has demonstrated a good understanding of the fundamental research questions. The use of hand-coded basis functions is appropriate for the initial stages of investigation, but other alternatives could be considered that have a basis in coding theory—for example, the use of low-density parity check codes. This research thrust is at a preliminary stage, and it is too early to decide whether an accelerated resource track would result in a high-impact transfer.

A research effort related to large-scale network experimental data reduction has led to acceleration of processing time for measured network packet traces. The approach uses a standard relational database with a combination of custom query processing that includes map-reduce and data summarization and compression. Providing several new visualizations of network operation augments the approach. The work is of high value and is adequately funded. One possible extension of the work would be to advance the state of the art for visualization through the use of nonlinear manifold learning techniques and unsupervised explanation of system regimes to reveal anomalous phenomena.

The research portfolio includes a focus on large multiscale material modeling with a special emphasis on identification of damage modes. Molecular dynamics simulations have been used in conjunction with a sectioning data reduction approach to develop a new and elegantly simple hydrostatic criterion for spallation of crystalline materials. The new criterion was then used to visualize the stress state from the data set obtained for an impact simulation. The research provides a new characterization of spallation. Furthermore, the approach, using molecular scale simulation coupled with novel data characterization, is broadly applicable in related problems.

The work of the visual simulation laboratory is focused on the use of a visualization-based framework to allow users to steer a multiresolution PDE simulation. The principal contribution of this work is in the computational steering of a parallel algorithm, but it was not clear whether the innovation was in the parallelization approach or in the visual framework in which the simulation was performed.

Challenges and Opportunities

Currently there is no research roadmap that identifies specified projects and goals linking research efforts to the stated goals—the science of large data, computational mathematics for data analytics, and real-time data access and analytics. The portfolio of research is primarily focused in computational simulations on HPC platforms. The portfolio could be broadened considerably given ARL's varied and demanding real-world data intensive applications that go well beyond computational simulations. Some examples are big data analytics in the context of social networks, geographic information systems, and situation awareness (e.g., in the context of cybersecurity). There is an opportunity to take a leadership role in developing scalable numerical optimization methods and multiscale and linear algebra methods for state-of-the-art machine learning that are suitable across a wide range of architectures, including large HPC systems, tactical HPC, and mobile systems. Existing expertise and strengths in the areas of numerical analysis, scientific computing, and parallel computing need to be leveraged in this context.

The data sets of interest for Army applications are characterized by their complexity and high dimensionality, and computing constraints often dictate computations that are approximate or statistical in nature. The data-intensive sciences research could be strengthened by incorporating more rigorous problem formulations that take such constraints into account and that lead to results with theoretically grounded error bounds. The research would contribute to ARL focus on verification and validation in computational science and would also enhance visibility through improved quality of publications. Given that many applications will combine big data, HPC, and high-performance analytics, there is a tremendous opportunity for datacentric computing to impact the design of HPC systems and their operating environments.

ARL is a Department of Defense (DOD)-wide supplier of supercomputing capacity and networking. The state of the art has today shifted to an emphasis on data-rich computations. In many sciences, large, widely accessible databases—for example, protein databases and large language corpora—have played a fundamental role as engines for progress. Military-focused computations can benefit today from such an approach. For example, huge data sets comprising network data, textual interactions, and even audio logs are generated and archived by large training exercises. The collection, curation, and dissemination of such large military-focused data sets could be vital to the ARL research mission as well as future engineering and technology efforts throughout DOD. This is a natural evolution for ARL. Serving these data could be over the same communication networks that ARL currently maintains. The data could be housed in the same supercomputer centers that ARL operates for DOD. Ultimately, such a service would be far more valuable than just providing raw compute cycles.

It is important for ARL researchers to anticipate new technologies and how they could impact its research agenda. One example of a technological trend to monitor is the emerging disruptive revolution in memory technology. Akin to the effect that cheap high-volume, solid-state drives (SSDs) have had on the supercomputing landscape, a larger disruption is expected with the high-speed, large-volume nonvolatile memories that will be available in the near future. How this will affect architectures and especially algorithms and software design is unknown today. ARL could take a leadership role in understanding the impact in general and especially with regard to Army applications.

Similarly, ARL research could also consider a focus on the missing elements of a computational architecture that could provide battlefield information dominance architecture (IDA). ARL is contributing strongly to this goal in the front end, via advanced sensing and networking. For the middle layers of an IDA, where data are reduced, fused, correlated, analyzed, and reduced to actionable information, there is evidence that ARL is building leading capabilities and advancing the state of the art. However, it would be desirable for ARL to devote more attention to looking at the back end of the IDA, where information is delivered to the individual warfighter. An information chain is limited by its weakest link. Beyond human factors research, research into the optimal adaptation of new and emergent technologies required for architecting the back end is required. As an example, low-latency, high-fidelity virtual reality (VR) devices and rendering engines made possible by advances in computing and organic light-emitting diode (OLED) displays will reset fundamental user interface (UI) assumptions in a far more profound way than did two-dimensional (2D) graphical user interfaces (GUIs), mobile or touch. The implications of these devices for the back end of the IDA are equally profound. An opportunity exists for ARL to take a leadership role in understanding how advanced displays, visualization techniques, and algorithms can be used, beyond the current applications to training, to reach an integral part of the future battlefield.

PREDICTIVE SIMULATION SCIENCES

The research program in predictive simulation sciences reflects an appropriately broad understanding of the underlying science and of comparable R&D activities at other institutions and agencies. Many of the research projects include collaborations with academia, federal laboratories, and industrial partners, and a few already have transition plans in place to deploy the results of the research in appropriate national security settings. Most of the publications cited are of recent vintage, and many have academic or industrial collaborators. A review of the work shows a few gaps in broad understanding, largely explained by the nature of the work, which requires expertise in both computational sciences and in the application domain. This underscores the importance of collaborations in the work.

Numerical models are generally relevant to the work proposed. Many of the projects are currently compute-bound and are now exploring ways to more effectively utilize additional computational capabilities available via GPUs or other exascale technologies. Close collaboration with leading universities and federal laboratories actively working on exascale computational technologies needs to be encouraged in ARL's Computational Sciences efforts to ensure that state-of-the-art developments in these fields can be transferred to ARL's Computational Sciences R&D projects in a timely manner.

The research qualifications of the ARL staff are relevant to the technical challenges inherent in the R&D work. There are some indications that the cultural changes necessary to effect a transition to a campaign-based strategy for R&D initiatives are still in progress, but no substantial expertise gaps were observed in the predictive simulation sciences presentations and poster discussions.

The R&D efforts presented incorporated appropriate levels of theory, computation, and experiment. Experiment could use existing validation data to ensure that simulations are predictive, but some of the research utilized real-world laboratory results to steer or refine computational science predictions.

For example, the meso- and microscale weather project utilized airborne weather sensors to improve microscale models in near real time. Additional effort in integrating theory, computation, and experiment via formal validation and verification (V&V) techniques would be a welcome addition to many of these projects.

Some exceptionally promising projects found in the Computational Sciences Campaign R&D portfolio are listed separately below. Most projects showed evolutionary progress, but there are projects that may provide revolutionary benefits to the ARL mission.

Future outside-the-box, close collaborations between predictive simulation science R&D projects and information sciences efforts have the potential to create novel hybrid technologies from the fusion of these two technology arenas. Large-scale computational simulation and large-scale data analytics seem an especially appropriate fusion for ARL R&D investigations.

Accomplishments

Multiscale material modeling is a potentially game-changing computational technology for predictive simulation in the mechanical sciences. The work on multiscale simulations based on scale-bridging combines several important threads of cutting-edge research into a useful software product that has the potential to provide higher fidelity deterministic and nondeterministic forward simulations in fluid and solid mechanics. These multiscale simulations are essential for assessing vulnerability, lethality, and effectiveness of weapons and protection systems, and the current effort demonstrates the project's utility in theory and also in practical application to software commonly used by DOD (e.g., Lawrence Livermore National Laboratory's ALE3D production software tool,[3] used for high-explosive weapons and target simulations). In addition to this research being of high value for predictive forward analysis, its multiscale sampling subsystem can lead to vastly improved performance for inverse analyses and quantification of margins of uncertainty (QMU) estimations. The latter, in particular, are often computationally expensive for application to large-scale problems. The principal technical challenges for this work lie in the availability of HPC resources, because the smaller scale simulation components can consume prodigious numbers of computational cycles. The research group has worked to overcome this challenge by deploying an optimal sampling scheme to reduce computational burdens on the smaller scale simulation components and by utilizing software broker components to aid in the load-balancing required for optimal use of HPC resources.

An important focus of the ARL research related to developing predictive capabilities for use on low-power computer platforms representative of what would be available to Army personnel in the field. Research on synchronous time-driven simulation for HPC accelerators focused on the simulation of large ensembles of random events for tactical cybersecurity applications. The research approach was to simulate the continuous real-world setting by utilizing a finite-state approximation that would converge to the continuous case as time step size decreased. Synchronization of events is problematic, with longer time intervals and smaller time step size, causing computational costs to increase significantly. The focus is to seek statistical convergence for the larger time steps. Validation of results on larger and more realistic data sets would be warranted.

In the area of materials modeling, research on scalable algorithms for simulating dislocations in microstructured crystals focuses on fusion of discrete dislocation theory coupled with continuum finite-element approximation, toward the capture of inelastic effects in polycrystalline and heterogeneous mate-

[3] ALE3D is a 2D and 3D multiphysics numerical simulation software tool using arbitrary Lagrangian-Eulerian (ALE) techniques.

rial structures. The research is promising and has broad applicability for material and structural failure simulations. Each element of this algorithmic fusion has strengths and limitations, but the coupling of these two techniques promises to remedy the weaknesses of each while preserving their considerable strengths. One of the most difficult technical challenges for this work is the effective load balancing of HPC resources. In computational physics, localization phenomena are not handled effectively through static load-balancing techniques, and the research could benefit from consideration of new developments in adaptive parallel load-balancing techniques. Because a similar challenge arises when transitioning from current computing architectures to emerging pervasively threaded hardware such as GPUs, due attention to research and development efforts in academia and other federal research institutions efforts is warranted.

The research on computational drug discovery focuses on developing efficient techniques for discovering which readily available molecules may be suitable for synthesizing drugs for use against pathogens that an adversary might deploy as biological weapons. This effort leverages a variety of university collaborators and is helped by access to ARL HPC resources such as the Excalibur supercomputer. The broad impact of the success of this research is significant for national security and in public health settings. Considerations relevant to technical limitations of this work include whether currently available libraries contain potential drug candidates, whether the identified compounds are optimal in effectiveness, and whether limiting the scope of these searches to small compounds is the best path for drug discovery.

Challenges and Opportunities

Predictive simulation sciences are increasingly important in DOD R&D venues. A broader understanding of the strengths and weaknesses of modeling and simulation techniques and of their impact on Army R&D practices is essential to help steer ARL research activities toward full realization of future simulation trends.

A challenge that cuts across the predictive simulation sciences portfolio is the lack of cutting-edge R&D efforts in validation, verification, and uncertainty quantification. V&V are indeed essential for all simulation applications, and failure to attend to these basic measures puts all of ARL's computational effort at risk of being of little value in the decision-making process. It is therefore important to raise V&V and related research in QMU as an important intellectual thrust in the ARL campaigns.

Many of the presentations and posters referred to an interest in utilizing emerging hybrid (i.e., thread-scalable with distributed memory, commonly termed MPI+X) computational architectures. The researchers at ARL who work in these intellectual venues can learn much by leveraging the considerable ongoing R&D efforts in advanced programming models for exascale computing that are occurring in leading universities and federal laboratories. In addition to these promising HPC developments, many of the computing-at-the-tactical-edge projects could also benefit from closely following these ongoing R&D efforts. One promising side benefit of exascale HPC research is the development of lower-power petascale and terascale resources that would be appropriate for this class of tactical computing.

The use of computational fluid dynamics (CFD) tools for the aerodynamic analysis and prediction of flight behavior of Army ballistic weapons parallels similar work in academic, industry, and federal laboratory settings. The coupled aerostructural simulation tools provide useful preliminary analyses at low Mach numbers. A serious limitation of the ongoing work is the assumption of rigid-body response for the projectile. While such an assumption would greatly simplify the computation, its validity has to be questioned for slender projectiles. Furthermore, while the assumption of rigid-body dynamics would hold for conventional artillery projectiles, finned structures are often sufficiently flexible and would impugn the accuracy of the predictions. Additional effort needs to be directed at extending the methodology for larger scale problems, higher resolution analyses, and optimizing the computational codes to take full advantage of emerging computational architectures.

OVERALL QUALITY OF THE WORK

There has been a significant improvement in quality of the research over the past 3 years, as indicated by the methodology pursued, the quality of the personnel assembled, and the infrastructure capabilities that have been developed over this period. The majority of the research presented (in both scheduled presentations and in the posters) was of comparable quality to that conducted in leading research universities or in federal and industrial laboratories. There were some projects where researchers clearly demonstrated a deeper knowledge of existing theory, of prior research in the field, and of state-of-the-art technology in relevant areas. Other projects provided a clear problem statement, including the articulation of a specific strategic direction for the research. It would be important to mentor researchers so that these best practices are embraced more broadly across all research groups. In each R&D effort, due emphasis was given to publication in relevant journals or proceedings, including (but not limited to) the open literature. Collaborations with university researchers via the ARL open campus initiative would provide opportunities for ARL staff to gain experience in the professional value of a continuous publication record.

Over the next decade, big data, complex analytics, and modeling and simulation will all come together in critical ARL applications and environments. This is evident from the materials provided by ARL in the area of computing sciences, data-intensive sciences, and predictive simulation sciences. Many of these applications will execute in a battlefield environment, where security and resilience will be of paramount importance.

To gain transformational improvements in power, performance, and resilience, the entire stack—hardware, software, and applications—needs to be codesigned. ARL has a tremendous opportunity to be a leader, especially in battlefield environments. To accomplish this, ARL would benefit from tasking an enhanced team of technical leaders to ensure that the subareas are increasingly engaged in real codesign.

The Computational Science Campaign is an inherently interdisciplinary enterprise, providing enabling technologies that cut across all of the ARL R&D campaigns. The capabilities of the Computational Sciences Campaign support virtually the entire spectrum of ARL R&D efforts. A broad swath of the projects could benefit from a better understanding of the theoretical and mathematical foundation that governs these problems. An increased contact and collaboration with researchers strong in theory—for example, from mathematics and computer science as well as the foundational science domains—would help achieve this. In particular, adding more computer scientists to the computational sciences research effort would encourage a more rigorous approach and would help advance the research agenda.

The projects reviewed comprised a mix of long-range basic science research and short-range research addressing tactical challenges and opportunities. The short-range research work is important, given the many challenges faced by the Army in its fields of engagement; ARL is developing innovative solutions to these high-priority tactical problems. It is also critically important that ARL continue to bring greater emphasis in its research portfolio to long-range strategic research in advancing fundamental knowledge in the underlying science. Strong relationships with key external entities have proven to be valuable across the breadth and depth of ARL research activities. This is an important positive development for ARL, and engagement with additional leading research institutions would further strengthen the awareness and capabilities of ARL researchers in important strategic areas.

6

Sciences for Maneuver

The Panel on Mechanical Science and Engineering at the Army Research Laboratory (ARL) conducted its review of ARL's vehicle intelligence (VI) programs—human–robot interaction, intelligence and control, and perception—at Adelphi, Maryland, on July 8-10, 2015; and ARL's vehicle technology (VT) programs—platform mechanics, energy and propulsion, and logistics and sustainability—at Aberdeen, Maryland, on June 28-30, 2016. This chapter evaluates that work, recognizing that it represents only a portion of ARL's overall Sciences for Maneuver Campaign.

VEHICLE INTELLIGENCE

Human–Robot Interaction

The human–robot interaction (HRI) research at ARL reviewed by ARLTAB in 2015 was of top quality. The researchers presented studies with rigorous design, evaluation, and analysis, and all used appropriate metrics in their evaluations. They have a thorough understanding of related research and have built collaborations with others at ARL across campaigns and have connected with the right faculty and laboratories in academia through the Robotics Collaborative Technology Alliance (RCTA). The research team has benefitted from an expansion in the postdoctoral program and early-career hiring, which has grown the capabilities and collaborations of the team.

Each individual project defines its own appropriate algorithmic or experimental metrics, but the researchers did not uniformly communicate the higher-level success criteria—that is, what makes their project successful in the context of the larger HRI effort. Such metrics need to be defined at a programmatic level and also be communicated well to the research team; some researchers could not identify the larger goals of the program.

The science in the HRI program is technically sound, and the work is published in top journals, including *Human Factors*. The work needs to have broad exposure, which is achieved through presentations at conferences and meetings. The utility of the work appears to be recognized within ARL—for example, elements from the tactile feedback project will be incorporated into the next warrior experiment.

The use of soldiers in experiments is commended. The move toward more realistic warfighting vignettes and more real-life simulations that instantiate threats and hostile elements would help to establish the value of a technology in achieving a desired capability.

Some researchers were embedded directly with the soldiers for a few weeks, and their experiences directly led to the formulation of research topics. This type of exchange is of great value for the research program. Also, some researchers are permanently stationed at U.S. Army installations, where they have regular contact with active-duty soldiers and are able to recruit soldiers as participants in research.

The research presented will be shifting from one-person/one-robot studies to multiperson/multirobot scenarios. This shift in focus is appropriate as the Army moves to the use of more complex teaming architectures. This new direction will bring a need for additional research in trust and in HRI, raising questions about how one verifies software and validates systems to build confidence in the joint human–robot system, considering complex, emergent behaviors. This use case also highlights the importance of providing the right information at the right time to the humans and to the robots, identified as a thrust of the HRI program.

Human–Robot Trust

In recognition of the changing role of people in a human–robot system, this topic investigates the relationship between trust and the design of vehicle autonomy. The goal is to look at design factors proactively in simulation concurrently with technology development, rather than retroactively; the design of human–robot interaction during early system development is important in order to create a robot system that will be used effectively and as intended. This project is commendable in that it is conducting evaluations and experimentation concurrently based on the applied robotics for installation and base operations system development at the U.S. Army Tank Automotive Research, Development and Engineering Center (TARDEC). The simulation models are based on real data from the system under development at TARDEC.

The milestones are clearly defined, appropriate, and feasible. This 6.2-funded work (early applied research) may lead to shorter-term application than the 6.1-funded work (basic research) described for most of the other HRI projects, though both may have longer-term implications. Trying to visualize what will be available in the way of automated vehicles in the longer term is challenging. This work seems to be a good stepping-stone to future interactions of humans with autonomous vehicles. Additionally, the close interaction with TARDEC on this project shows that the research is targeting a current Army need.

The researchers turned up more than 300 definitions of the simple-sounding word *trust*, a fundamental variable when working with people. They explained convincingly that trust is a critical factor to be considered when studying the interactions between humans and autonomous vehicles.

A three-factor model for trust, developed for the RCTA, served as the foundation of the researcher's Ph.D. thesis on human–robot trust. The effort takes an analytic and empirical approach, deconstructing the factors involved in trust, into three categories: human, environmental, and robotic. One known factor that influences trust is system reliability. The effects of many other factors, including stress, workload, personality, trust propensity, and coping style, are still unknown. Quantifying this space is the basis and motivation for the project.

The equipment and tools that are being used are appropriate for this early investigation. The team is in collaboration with TARDEC as well as collaboration across ARL directorates, using the Control of Autonomous Robotic Vehicle Experiments (CARVE) and the Robotic Interactive Visualization Toolbox (RIVET) simulation tools. The trust theory was developed through the RCTA. The autonomous vehicle is under development by TARDEC, and current work uses computer-based simulation. The simulation is low fidelity—a reasonable first step for an early investigation. The simulation is based on data from the real system at TARDEC, and the simulation is updated as the actual system is developed further. Improvements in the simulation's impact could be achieved with some affordable upgrades to the current desktop solution: for example, with wrap-around hardware.

There is some question about the effect of perception of risk on the experimental results, given the use of a simulated environment instead of a real one. Prior research in HRI has shown that the perception of risk, or lack thereof, influences the behavior of participants in studies with human subjects.

The work has been published in top journals, including *Human Factors*. Broad exposure of ARL's work is necessary and is achieved through presentations at conferences and meetings.

The findings on how different people trust an autonomous system illustrate the complexity of the challenges faced by the new technologies. The project also demonstrates the effectiveness of the RCTA mechanism for conducting Army-relevant research in academic environments and the pathways that are created to hire researchers into ARL.

Multimodal Displays for Human–Robot Interaction

The motivation for the project is well-founded: Indeed, a soldier's visual modality is overloaded. Additionally, constraints on auditory communication may pose a threat to soldier security and mission success. This project investigates the use of other modalities for communication with the soldier, including tactile and gesture. The project goal is to identify interaction modes (single or in combination) that are most effective for soldier-robot communication. The focus is on achieving external validity by synthesizing and translating theory-based predictions, meta-analysis, and experiments to determine whether results generalize to the field.

The project is commendable in that it conducted experiments with real soldiers in a realistic environment and context. The principal investigator is an outstanding practical researcher, and the effort is a clear success. The work was disseminated well at conferences and meetings; journal publications would be very valuable to the research community.

The experiments conducted would have been improved if they had explicitly included in the scenarios a robot as a critical element to provide meaningful information to the soldier, rather than focusing on the communication medium itself. Additionally, the information passed to the tactile belt could have been provided as easily from a human as from a robot, and so other scenarios involving human–human communication using the tactile belt could be performed.

The comparison of text messages, which remain on the screen, versus one-time tactile input, is questionable. It would be reasonable to allow the soldier to re-trigger the signal, which would provide a better comparison of the two methods in terms of the length of time for which the information is available.

Understanding how the technology supports the soldier in abnormal conditions is important, because there may be trade-offs in task performance and situation awareness. Additional insight into the effect of the tactile vest on the soldier's behavior is expected if disruptive scenarios are incorporated into the experiments.

Overall, the project has developed a piece of technology that has the potential to improve soldier performance, particularly in terms of safely passing information in combat situations.

Human-Autonomy Sensor Fusion for Rapid Object Detection

The project conducts research on models of fusion between computer vision and neurophysiological responses. Recognition systems do not integrate humans explicitly; this project looks at an alternative architecture in which humans are peers to share the information with autonomous systems. The approach fuses human neurophysiological response to enable rapid real-time target detection. The objective of the work is to evaluate the hypothesis that joint human–autonomous system target recognition surpasses the target recognition capability of the autonomous system alone.

The project defined appropriate algorithmic and experimental metrics. The larger milestones and success criteria were less clear, although the project is part of a larger project with 40 researchers.

The project addresses challenging technical issues; neural classifiers are not yet fully understood. The push-button response (part of the human target recognition component) lags behind computer vision, and the neurophysiological recognition lag depends on human conditions. The goal is to produce overall better accuracy and efficiency with a human sensor and to demonstrate a path forward to relevant target tracking and engagement scenarios.

At this preliminary point, the results support the hypothesis that the human component improves object detection, but it is unclear whether the improvements are significant from a practical standpoint. While the research is young and the delta improvement is small in terms of making progress to a fielded, capable system, the payoff could be significant. One manuscript has been submitted for a journal publication and another was accepted and presented at the 2015 International Conference on Intelligent Robots and Systems. This project is currently limited to object detection, but it could be expanded to explore similar approaches for object classification.

Intelligence and Control

The intelligence and control (I&C) research theme is focused on enabling the teaming of autonomous systems with soldiers. The overarching goal is to develop an intelligent autonomous system of robots that can perform effectively in an uncertain environment by optimally using limited resources. The long-term goal as articulated by ARL is to minimize the human presence in the battlefield. The topics investigated in this research area are aimed at generating the technologies needed to meet the challenges posed by future military operations that could occur in military-relevant missions and in military-relevant environments. To meet the goals of the research theme, six research programs have been initiated:

- *Control* focuses on the low-level processes and closely couples sensing and action (actuation) of individual elements of the vehicle;
- *Planning and guidance* focus on the mid-level, vehicle-centric layer of the control architecture, with immediate path-planning objectives;
- *Abstract reasoning* focuses on the cognitive element of the architecture, with a special emphasis on human–robot teaming occurring at this level of the architecture;
- *Teaming and coordination* focus on the interaction of multiple homogeneous or heterogeneous entities to achieve a specified goal, including coordination and communication;
- *Behaviors* focus on actions of a vehicle built from a hierarchy of elemental tasks and capabilities to achieve one or more specified goals; and
- *Learning and adaptation* focus on employing key cognitive features of an intelligent vehicle to enable control of its actions and/or behavior to successfully achieve goals in dynamic and/or unrecognizable scenarios and environments.

Projects in the teaming and coordination and learning and adaptation project areas were not presented for review.

The focus of the I&C theme is developing software and algorithms that enable the vehicle to approach a higher level of cognition, enabling the teaming of autonomous systems and soldiers. The I&C theme has tight couplings with the RCTA program and the Micro Autonomous Systems Technologies (MAST) Collaborative Technology Alliance program, with some specific ARL focuses. The higher level cognition that the I&C theme focuses on is aimed at enabling autonomous assets to work in environments of relevance to the military—caves, subterranean spaces, jungles, undercanopies, megacities, and urban environments. Specific I&C research topics are targeted to leveraging state-of-the-art approaches and expanding them to address the uniqueness of the military environment and missions. To identify specific research projects that address the theme, the Army needs, which emphasize high-level capabilities, are deconstructed into specific project areas. The following projects were presented for review.

Abstract Reasoning: Spatial Reasoning in Uncertain Conditions

The focus of this research was on determining how to characterize information collected from field data represented as the collection of uncertain (or incomplete) information. The primary issues this research is trying to resolve are how to build an initial knowledge base and how to expand it autonomously when needed. The state of the art in this domain has been achieved by Microsoft (Bing), Google (Knowledge Graph), and IBM (Watson). The primary limitation of current approaches is that they are trying to extract all information prior to deployment (versus at query time). Overall, the principal investigator (PI) demonstrated broad understanding of the field and possesses the skills necessary to perform this research. The PI employs adequate tools and methods for this research and has several papers published, including in well-respected venues such as the Association for the Advancement of Artificial Intelligence.

Opportunities

This is an analytical study that is still a work in progress. It was not completely linked to specific military-relevant applications and did not involve any real-world experiments. Issues that might impact the success of the work include how to determine whether outcomes are reliable and correct; how to get algorithms and operations to run on the robot processor (along with the other processing components); and understanding key elements of the development path necessary to get to that point. Only limited results were presented. Estimates of the algorithm accuracy were not provided; more effort is required to provide evidence of feasibility. Experiments would also be helpful. Growing the PI's team might help in addressing these challenges.

Planning and Guidance: Autonomous Mobile Robot Exploration with an Information-Gain Metric

This research applied an information-based approach to the mobile robot exploration problem, based on probabilistic and entropy concepts. Effective robot exploration is important for intelligence, surveillance, and reconnaissance efforts. It was shown that the developed algorithm could be used to more effectively detect improvised explosive devices (IEDs). Probabilities or weighted probabilities could also be used to reflect prior information. As such, the problem being addressed is particularly relevant to the mission of enhancing the warfighter's capabilities via intelligent robotic teams. The algorithm presented was compared to a baseline (non-information based) algorithm. The PI was well aware of the state of the art in this domain and is well qualified to pursue these important issues. The PI's collaborative efforts, which led to a number of publications, are excellent.

Opportunities

While the information-based entropy algorithm was shown to outperform the baseline (greedy) algorithm, work needs to focus on describing the pros and cons of each algorithm, comparing the algorithms presented to other algorithms considered in the literature, examining the effect of local topology on the algorithm's temporal and spatial performance, examining the effect of a priori probabilities and weightings on algorithm performance, its reflection of real-world concerns (e.g., uncertain communications, terminal hazards), and extending the work to include multiple robots.

Behaviors: Scene Consistent Visual Saliency

This research dealt with how to better define visual saliency in a dynamic environment. The project was first envisioned because ARL researchers were examining images from a moving robot and realized that present definitions of saliency did not give consistent metrics for saliency—for any given feature—as the moving camera passed a feature and dynamically changed its field of view. The goal was to define visual saliency in a manner that was independent of the view in a general way. As a step toward this, the research considered images obtained from a fixed camera as it panned, tilted, and zoomed to give different views. This was the first step toward defining a more general definition for a camera with a moving base. The approach began with a bottom-up method of determining saliency, in which one looks for whatever jumps out of the background. Later, this approach was merged with the top-down approach of looking for specific objects. Using a consistency metric, it was realized that consistency for the saliency of a given feature in the surroundings can be obtained only if various images (from different views) are merged into a composite mosaic that casts the set of pixels as a unified field of view. From this concept came a consistent definition of saliency of the visual image. The project is well thought out and has demonstrated that the new definition of saliency can give consistent answers for saliency over a range of camera angles and zoom parameters. The scientific quality of the work is excellent, and the researchers are qualified to do the work. The researchers understand the positioning of this work with respect to other work in the field and have published in appropriate journals and conference proceedings.

Opportunities

The approach represents a first step in addressing a larger problem. It does not work well when the focus is changing, so additional improvements in methodology are needed. This work could also be improved by use of a depth parameter via the stereoscopic effect (once the base is allowed to move). Researchers could then better understand how the work connects with other human–robot interaction work in ARL. The researchers need to identify how the work addresses an important need—that is, why is it necessary to measure saliency consistently and how good does saliency consistency need to be for images to be useful for the Army mission? An additional opportunity could include integrating eye-tracking devices to try to determine what a solider thinks is salient, inferring what the solider is trying to convey based on eye-gaze, and providing that information to the robot or informing the robot planning algorithms.

Control: Autonomous Self-Righting for a Generic Robot with Dynamic Maneuvers

In this research, the problem of self-righting a fallen robot was examined. This situation occurred when the robot fell despite all that had been done to prevent toppling, so the problem is an important one. A potential energy method was used to address the problem. The method is based on an approximation approach that was shown to be potentially useful for robot design, minimizing the number of states from which acceptable recovery is not feasible. The PI has provided metrics and validations to evaluate the accuracy of the approximation. The associated publications of the PI are also good, and a related patent has been filed.

Opportunities

The work could examine how well the energy approach taken by ARL works in practice and a more precise (albeit computationally more burdensome) measure of acceptable recovery (e.g., a measure that includes spatial/temporal constraints). Although the focus is on righting the robot from a static overturned position, a better understanding of the dynamics would come from a more general examination of a broader set of related issues, including stabilizing and destabilizing factors before overturn and the ability to right itself while falling and rolling before coming to rest.

Behaviors: Autonomous Navigation Analysis

This project had the objective of evaluating the performance of three different Army robots navigating from a fixed starting point to a number of predetermined global positioning system way points in an outdoor environment. The evaluation was at the system level—that is, the overall performance of the robots, including their sensors, controllers, and traction/power train systems, was included. The project was designed to expand on research that had been conducted indoors by Microsoft and apply it to the more Army-relevant situation occurring outdoors, in unfamiliar surroundings and with various obstacles that cause significant difficulties for the robots. The experiment was conducted at the Aberdeen Proving Ground facility, and the errors, difficulties, time, and success rates of the robots were evaluated. Further experiments are planned over the next year to enrich the comparison of system performances. The referenced work done by Microsoft was an appropriate starting point. The researchers were qualified to perform this research and they conducted the research as a team and used state-of-the-art facilities to conduct their analyses.

Opportunities

This project was not a sophisticated research effort, but it was a good start in evaluating the performance of a system. Future work could focus on developing methods to predict system-level failures and understanding what causes them. It could conduct more detailed experimentation and analysis to better grasp the effect of the various system components on the robot's performance.

Abstract Reasoning: Robotic Dream—Episodic Memory Consolidation and Revision

This research focused on the development of a memory system that allowed a robot to retain knowledge from previous experiences. There was some uncertainty about how long this particular project has been pursued. It was listed as a 2004-2020 effort, but it was unclear whether the project was ending or continuing through 2020. There was mention of a path forward, including transition to CTA partners, although this was not defined. It would be helpful if the objectives included more metrics-oriented information and were more clearly set forth.

Storing and processing only exciting events and precursors is an elegant and logical contribution, and there is also a process to code the events into a set of simple low-memory symbols. Memory access time savings also seem intuitive, and they are dramatic.

One PI indicated that no similar research was being done elsewhere, but this was belied by the numerous references cited in a recent journal paper of the PI, which had a sufficient scholarly section of related research. The PIs could continue to strive to publish work in mainstream journals in the field, such as those of the Association for Computing Machinery or the Institute of Electrical and Electronics Engineers.

Opportunities

Since this research has been in existence for a while, there needs to be a stronger linkage, at this point, to military-relevant scenarios.

Overall Accomplishments

The I&C team is developing and supporting a suite of forward-looking technologies, algorithms, tools, collaborations—all of which are important for the warfighter effort. ARL has brought together a number of different viewpoints and different skills from different disciplines to start thinking about these problems and tackling them in innovative ways. Collaboration with universities and other agencies and industry has been active and of high value, and ARL has invested in the research through the people it has hired. In general, PIs recognize related research and methods and leverage them to push forward improvements to address the uniqueness of military-relevant problems. While unifying demonstrations, milestones, objectives, and capabilities could better motivate the specifics being developed and elucidate how they will be integrated, the developments being pursued are for the most part essential.

Overall Challenges and Opportunities

The I&C team needs to move on to the next step—bringing the different research projects together synergistically to successfully address broader problems faced by the Army. The I&C team has a good start on this but needs to formalize the process for integrating and knitting together the various research pieces necessary to transition to the next step. Developing quantitative milestones for gauging progress and performance would help in this endeavor. By using milestones, research projects can be redirected where appropriate to better achieve the overall mission goals. It would also be good to see the process whereby desired future capabilities or goals are broken down into a sequence of achievable (realistic) short-term capabilities and goals.

There also needs to be a focus on the big picture: an understanding of where ARL is going and how the projects fit into the bigger picture. To quantify general progress and application-specific performance, more specific connections to the literature could be made in the course of baselining or benchmarking. All projects could make sure to reference the related literature (baselines, metrics, or benchmarks) and understand how they fit and compare. Through mentoring, guidance, and appropriate milestones, quicker progress might also be made toward integrating projects and more effectively contributing to solutions to important problems. Mentoring can come from both internal and external experts. ARL's open campus concept can be used to bring in external experts.

Challenges to I&C research focus on determining (1) how to deal with trade-offs in order to determine which research to continue; (2) how to effectively integrate outcomes from the individual projects and develop a methodology for this integration; (3) how to share the overarching systems perspective and relay that vision to the research projects; (4) how to identify and validate the process of getting from high-level capability or needs to research tasks (and evaluation or benchmarking of whether they comply with needs); (5) how to appropriately delineate between basic and early applied work; (6) how to balance and integrate top-down and bottom-up-driven processes; (7) how to compare the research against the standard baseline data sets (when available) and how to identify standard metrics for validating whether the research has achieved the stated goals of the proposed work; and (8) how to transition the research from work on simplified problems that facilitate analysis to actual scenarios that are germane to the Army's unique problems and characteristics.

Perception

The ARL aspires to be the nation's premier laboratory for land forces. The perception group at ARL has made significant headway toward achievement of this goal. It has succeeded in establishing relationships with top university laboratories, has attracted some outstanding personnel, especially new Ph.D.'s, and is undertaking interesting and relevant work on a par with academic departments. The ARL perception group is well aware of the current trends in the research community through participation in top, highly competitive conferences in the field. ARL's open campus policy appears to be making a positive difference in the quality of the work.

The perception group did not articulate clearly the context within which it works and did not clearly differentiate its approach from the approaches in other areas. Research work on robotics perception needs to be linked to the power needed for the robot or the materials of which the robot is constructed. A systems approach is needed here. Therefore, based on the material presented, the following were inferred:

- The perception group, for the purposes of this evaluation, works in the context of three scenarios:
  - Microautonomous systems and technology (MAST), where soldiers use microrobots to explore;
  - Robotics collaborative technology alliance (RCTA), where soldiers use robot/human teams to penetrate the built environment; and
  - Applied robotics for installations and base operations, where robots perform functions on bases, relieving the warfighters of these functions.
- Most of the perception group's efforts are basic research in one of the three scenarios. The goal of each is as follows: Within the next 5 years, perform relevant basic science; 5 to 10 years from now, inspire concrete modules for the three integrated scenarios; and 10 to 15 years from now, apply these modules to support integrated experiments, which will subsequently enter the Army Research, Development and Engineering Center as processes to be matured by 2040.

The perception research was assessed according to its achievement of the above-inferred goals.

Autonomous Squad Member

This project attempts to detect changes in the tactical situation by observing the motion of individual soldiers. It is an interesting and important problem as the Army integrates robots into small unit operations. It was not made clear whether there is a roadmap from the broad concept to achievable steps in that direction. Is it going to be practically feasible to deploy a robot with sufficient capabilities in the next 10 to 15 years? Will the robot participate beyond being a team member? In 2040, a robot may have unique capabilities—will it then be a more active participant in activities?

Data-Driven Learning and Semantic Perception

The core of this work is using video analysis to categorize the actions a human is performing using machine learning with a hierarchy of action templates. This is good, focused work using state-of-the-art methods. Like similar methods, there are still questions about its broader applicability. It would be informative to learn how this work might relate to or affect other projects in the group.

Efficient Discovery and Labeling of Environments for Visual Classification and Autonomous Navigation

This project is also using machine learning methods to examine video data. In this case, the intent is to segment the scene into regions such as trafficable areas, vegetation, buildings, and sky. This is a good combination of theory and practical application and has great potential. The work could become a framework for developing and testing other new ideas and scenarios.

Dynamic Belief Fusion

Dempster-Shafer theory is a well-known framework for reasoning about uncertain events when, because the categories might overlap, the sum of probabilities need not add up to one. The method introduced here updates Dempster-Shafer for object detection. The results indicate that the new method outperforms existing methods, but the theoretical basis has not yet been fully explored.

Immersive Display of Robot Lidar Imagery

ARL has its own design for a lidar that has excellent performance, generating 256 × 128 pixel depth images. This project takes that three-dimensional (3D) data and displays it on an Oculus Rift head-mounted display for visualization. The project is building the tools needed to enable future research efforts.

Real-Time Optical Flow

The basis of optical flow—measuring the apparent motion of a scene as the observer or the object moves—has been established for many decades. Typical optical flow techniques break down with large changes in illumination (e.g., the sun going behind a cloud) or with cluttered scenes (e.g., trees blowing in the wind). The work shown here demonstrates a new approach, which is much less sensitive to the absolute brightness of the scene and is capable of differentiating the optical flow from humans moving in the scene from the flow caused by vegetation. While the parts of the algorithm are not novel, the combination of these subunits shows promise.

Overall Accomplishments

The perception group presented a broad spectrum of projects, including understanding group behavior, perception for mapping and navigation, and basic research on dimensional reduction and clustering. Many of these projects are of high quality and are informed by the goals and needs of today's Army. The projects are well defined, and the researchers are aware of the state-of-the-art of computer vision. The group actively participates in important international conferences, which guarantees their awareness of the current and relevant activities in computer vision. Most of the work focuses on specific scenarios relevant to the Army. While it may not be as broad or groundbreaking as the best university research, the work is appropriate to the ARL context.

Through the MAST and RCTA collaborations, the perception group has access to many of the best computer vision researchers in the country, including those at University of Illinois; University of California, Berkeley; University of Southern California; John Hopkins University; and Carnegie Mellon University. Although these collaborative projects at the universities were not within the scope of this review, it seemed clear that they are a good chance to enhance the visibility and quality of the science at ARL. These collaborations need to be exploited to continue to strengthen the group.

The facilities, equipment, and approaches are well-targeted at the group's goals. The laboratories appear to have created a good computing and experimental environment, and the test site at Fort Indian Town Gap provides realistic scenarios for testing against the three scenarios.

The researchers seemed to work smoothly together, with encouragement from their management to cross organizational boundaries both within ARL and with other national laboratories. At the same time, it was not apparent what mechanisms exist to encourage sharing information; a regular seminar series might be a good addition.

While the quality of the work overall is very good, there was no single project that stood out as especially promising and ready for accelerated deployment. The closest is the work on weakly supervised segmentation for mobility. This project is significant for several reasons: an interesting vision/science result was published at a major conference; it is an integrated end-to-end project that demonstrates the value of the research; and it is an external collaboration with a university. These are all indicators of the project's scientific value and intrinsic contributions. The more projects that exhibit those characteristics, the stronger the perception effort will be.

It would have been helpful to present, in a concise form, a list of publications, awards, and other data that would help to convey the recognized accomplishments of the group.

Overall Challenges and Opportunities

Perception is a rapidly evolving area. The tool sets, the approaches, and the benchmarks on performance change yearly (sometimes monthly). It is also an area in which it is highly competitive to hire, placing a premium on providing the best resources, colleagues, and opportunities to attract the best people.

The ARL has done a very good job of building a strong perception group and giving them the opportunity to perform very credible single-investigator basic research. To move to the next level, ARL needs to think about a few audacious challenges that go beyond the extant state of the art—so-called grand challenges. These challenges would provide an exciting context for the group and would provide a point of focus for collaborative efforts. This is not to say that the work might not continue to be basic research. Semantic labeling of scenes, for example, would contribute both to the various integrated scenarios and to the international perception community.

More domain-specific challenges for teams, such as off-road mobility, would speak to the broader ARL mission. There are many other possibilities, but it is important for the perception group to aspire to one or more audacious projects that speak to Army needs.

Another opportunity would be to pursue some common platform/sandbox concepts that could be built upon. For example, the project related to weakly supervised segmentation deployed on a robot could be driven in a wide variety of directions. Its current application for driving is already interesting, but perhaps that could be expanded to become a platform for testing other segmentation methods (e.g., motion-based segmentation) and human body tracking for teaming. Again, this would become a point of cohesion for the group.

Most of the work presented focused on relatively traditional RGB vision.[1] However, there is no reason to limit the activities to vision. It may be relevant to consider multispectral sensing, range sensing, and contact sensors, such as temperature, force, and pressure.

VEHICLE TECHNOLOGIES

Platform Mechanics

The platform mechanics area focuses on fundamental research to enable highly maneuverable high-speed air and ground vehicle platforms and subsystems for the future Army, ranging from large combat/cargo vehicles to microscale devices. Specific research programs include fluids, structures and dynamics (rotor dynamics), actuation and mechanisms (adaptive wing span mechanisms), and platform configuration concepts (capability assessment and trade-off environment). The key campaign initiative (KCI) addresses small unmanned system airfoil research and rotorcraft aeroelasticity research within the discover and advance vertical takeoff and landing innovations, novel, concepts, and ideas (DAVINCI) program. The core campaign enabler addresses mechanics and dynamics of complex systems with research on dynamics of fluids and structures interactions, morphing structures, and immersive, interactive technology impacts and trade space exploration.

Koopman Decomposition of Periodically Excited Hopf Bifurcation Systems

This is excellent work, bringing modern ideas from nonlinear dynamics into the Army set of tools. The applications will give insights into a number of areas, such as dynamic stall. Publications are strong with a good paper in the *Physical Review Journal*. Another great area of application would be the fundamental nonlinearities of ground resonance. The researchers might consider this as a possibility. The approach could very well be adapted to include methodologies for the design of control laws for realistic systems. This does not imply that a turnkey control software needs to be developed. It does mean, however, that the theory could be advanced to a state where, for example, a Ph.D. controls engineer at a helicopter company would be able to utilize the ideas of this group in the design of nonlinear control systems.

This project is identified as one of the outstanding presentations that was given at this review. The investigators have taken advanced mathematical methodology in nonlinear systems theory and applied those tools to current, meaningful Army problems. By the use of nonlinear modal decomposition, they have been able to take complex experimental data and decompose them into the fundamental modes of interest, thereby giving insight into the physics of the underlying mechanisms.

---

[1] RGB is an additive color model in which red, green, and blue light are mixed in various ways to reproduce a broad array of colors.

The fundamental mathematics used relies on the proof that for any nonlinear system there exists a mapping that represents its evolutions by an infinite dimensional linear system. Although it is not possible to perform exact computations on the infinite dimensional system, the researchers have shown that the original system and its nonlinear modes can nevertheless be approximated by a finite number of linear systems. This holds great promise for realistic improvements in understanding and in design.

This methodology can be used to give insight into the fundamental behavior of nonlinear systems. It can also provide a low-order model of such systems to be used in control-system design. The method is broadly applicable and not limited to one particular type of nonlinear problem. The investigators have used dynamic stall as an exemplar system upon which to exercise the methodology, and this example has generated keen novel insights.

The researchers might consider also looking into Floquet instabilities. The Army's need in their future vertical lift initiative to fly faster (and therefore to slow the rotors of aircraft) implies that these Floquet type of instabilities will become more and more important as one proceeds forward. Floquet instabilities usually are computed by periodic-coefficient equations that arise from linearization of a nonlinear set of equations about a periodic orbit. While classical Floquet theory typically neglects some of the nonlinear behavior, the present methodology might be able to properly account for nonlinear effects.

The researchers need to consider the possibility of applying the methodology to ground-resonance instabilities. These types of instabilities are so nonlinear that the military specifications (MIL specs) do not even call out a particular damping level for compliance. That would be meaningless for such a nonlinear system. Rather, the flight specifications declare that the limit cycle of the instability will be smaller than a certain amplitude of shaft wobble. This implies that the nonlinear methodologies studied here might be helpful. Ground resonance with one damper inoperative is also a crucial requirement for military aircraft. That condition gives the periodic coefficients of Floquet problems for which this approach could also help.

Other possible applications to periodic problems can exist even in the case of hover. For two-bladed systems, there are periodic coefficients even when there is no forward speed. Also, the interaction with wake dynamics (and the tendency of vortices to pair in hover) is another source of periodic nonlinearities that this nonlinear methodology could address. Again, the present approach could contribute. The researchers need to consider looking into how control system designers in the rotorcraft industry might be able to use the work of this group to design control systems for nonlinear flight systems.

Characterization of Coaxial Rotor Performance and Vibratory Loads

This work is long overdue. The effort could be both enhanced and accelerated to support the timely need for more fundamental coaxial-rotor methodologies required to support ongoing future of vertical lift activities. The project can do a better job in uncovering previous research that is directly applicable to the project, such as that available in the public literature on the Sikorsky XH-59 B ABC experimental aircraft and the Kamov family of coaxial rotors. This study of previous data seemed to be absent. The proposed plan for assessing the effect of rotor separation, rotor phasing, blade number, and so on is good.

While the work reported on the loads and vibration is applicable to the recent University of Texas, Austin, test efforts (and seemed to correlate well with computational fluid dynamics [CFD] work), it did not seem to explain the fundamental technical parameters applicable in the experiment such as rotor separation and hinge offset.

A more complete methodology could explore the results of stiff high-hinge offset rotors. Important parameters, such as rotor separation and upper and lower rotor phasing, were also absent from the presentation and ought to be explicitly considered. Furthermore, there was no effort to discuss the appropriate frequency placement of the target rotors. This could be a crucial element.

Possibly, the researchers were only concerned with the aerodynamics of the subject model rotors and not the necessary aeromechanics and dynamics. An additional direction for future work would be consideration of different blade geometries for the upper and lower rotors. Finally, the work still needs to be expanded to cover an anticipated range of advance ratios up to speeds of 300 knots. Eventually, the interaction of the rotor with the aircraft tail at various speeds could be explored.

Experimental and Analytical Investigation of Continuous Trailing-Edge Flap Actuation Authority

This project aims to apply state-of-the-art CFD methodology (HELIOS) to prediction of coaxial rotor aeromechanics. Specifically, rotor-to-rotor interactions (lower rotor affected by upper rotor downwash) and wake-to-wake interactions are studied. Questions being addressed are these: Can one exploit dissimilar rotors for improved performance? Can one use on blade systems to reduce vibrations? How can one improve accuracy and efficiency of prediction of vibration and deflections?

This is extremely important work that analytically and experimentally explores the ability of embedded blade flaps to achieve rotor collective and cyclic control using electrically stimulated trailing-edge elements. The work appears to be very well done with good supporting background that demonstrates an understanding of existing literature. Agreement of experimental voltage and displacement is very good. One of the findings is that, although adequate displacement for full cyclic control is achievable, full collective authority over the entire range of flight conditions is not yet achievable; and a relatively slow moving separate rotor pitch control—located within the drive shaft—will replace the current complex, mechanical swashplate and drag producing pitch rods.

The work could easily be expanded in two important ways. Specifically, the researcher might include an element of controls thinking at this early stage in the work. Any lack of controls thinking will complicate the future applicability of the work. Incorporation of controls thinking is necessary if one is to include plans for sensor-embedded, sensor feedback and, more importantly, if one is to include redundancy management. Additionally, the effort, if feasible, might be extended to explore the system response bandwidth appropriate to incorporate a modicum of higher harmonic control into the logic. This will probably entail incorporating a different trailing-edge actuation surface stiffness into the end solution.

Multi-Fidelity Modeling of Active Rotor Concepts

This project is focused on the development and validation of comprehensive dynamics and CFD models of rotors with active trailing-edge flaps. There is collaboration among ARL, Army Aero-Flight-Dynamics Directorate (AFDD), and Pennsylvania State University (in the form of a summer intern). The dynamics analyses utilized include RCAS (Rotorcraft Comprehensive Analysis System) and CAMRAD II,[2] and the CFD analysis is HELIOS.[3] The quantities of interest include hover and high-speed performance, aeroelastic stability, active controls, loads and vibration, and acoustics. The project seems to be proceeding well.

[2] CAMRAD II is an aeromechanical analysis of helicopters and rotorcraft that incorporates a combination of advanced technology, including multibody dynamics, nonlinear finite elements, structural dynamics, and rotorcraft aerodynamics.

[3] The Helicopter Overset Simulations software (HELIOS) is a multidisciplinary computational suite being developed by DoD and the U.S. Army. It includes modular software components for near-field and far-field CFD, off-body adaption, domain connectivity, rotorcraft comprehensive analysis, mesh motion and deformation, and an exact fluid structure interface module.

A conference paper was written this past year, and RCAS simulations of the Boeing Smart Materials Actuated Rotor Technology (SMART) rotor testing at NASA Ames Research Center was initiated. Ongoing tasks are making a good effort to work with HELIOS developers to move to the incorporation of that capability this coming year. When completed, this project will result in a very-high-fidelity validated simulation of the SMART rotor system. A conference paper has been written based on the initial stages of this work, and this is a very positive development.

It is suggested that the scope of the project be limited to rotors with active flaps, as opposed to numerous other active devices. It is also suggested that a careful review of previous validation and simulation efforts on the SMART rotor be conducted. More emphasis needs to be placed on an array of theoretical/experimental models to develop improved physical understanding.

Bio-Inspired Air Vehicle with Arbitrary Wing

This is a fascinating project aimed at the design of a system level performance model for RoboRaven—a robotic flapping wing drone. RoboRaven is an intermediate size drone designed to be able to hit the sweet spot of both maneuverability and speed based on programmed wing configuration and actuation. The wings are driven separately, and wing kinematic models were developed. The aerodynamic model—which calculates thrust, lift, and power—is integrated with the motor and battery model to provide a system-level model. This model is then used to create a performance map. A vortex-ring formulation is used for the system model. This avoids approaches that are based on low-fidelity coefficients as well as approaches that are computationally expensive, such as detailed CFD analyses.

The system-level model allows for an interactive design environment where trade-offs can be explicitly considered, such as wing size and motor weight. In addition to the modeling, experimental innovation includes the design and fabrication of an onboard sensor suite so that inflight data can be gathered in real time in realistic flight, as opposed to collecting data from a fixed stand in a wind tunnel. This aspect is important for realistic model verification for a flapping wing robot, which will be crucial to the success of the project.

The overall goal is to be able to use the developed and validated system models to evolve a perfectly adapted winged robot design for given applications. Overall, this is an excellent project that combines detailed and system-level modeling, meta-modeling, experimental developments, and characterization.

Axial Propulsion with Flapping and Rotating Wings

This is a very interesting area of investigation. It is an old-school analysis of the performance of rotating and flapping wings, which yields deep insight into flapping propulsion and how to best design a flapping system. In this study of axial propulsion by the use of either flapping or rotating wings, the investigators are drilling down to first principles in order to determine the set of parameters and the design space for which either flapping or rotating wings are the best alternative. In this light, they are utilizing the classical ideas of momentum theory going back to the work of Glauert in the 1930s. Although these concepts may appear at first glance to be crude in their assumptions, as compared with modern-day CFD methods, they nonetheless form an important piece of insight into how thrust and power are obtained when one grabs air particles and accelerates them to make lift and simultaneously produce drag.

The work has opportunities to be even more productive if the researchers will pay a little more attention to the literature, starting with Hermann Glauert and ending with recent work at AFDD by Mahendra Bhagwat. The importance of wake rotation, wake contraction, and momentum assumptions has been keenly investigated by Bhagwat's group in the past 5 years, building steadily on the work going back to Glauert and improving on those original assumptions. A study of that literature could give even more insight into the study of flapping versus rotating wings.

It is important that the researchers pay attention to their overall goals and have the wisdom to know when the investigation has taken them as far as it can. Once a certain amount of insight is obtained, it will be time to use that insight to proceed to more advanced methods, such as vortex-lattice, vortex particles, or CFD. The project might look ahead to determine how it will know when it is time to make that transition.

Performance and Controllability of Overlapped Quadrotor Vertical Takeoff and Landing

The objectives of this project have been to develop comprehensive tools to gain insight into the performance and aeromechanics of novel quadrotor configurations with overlapping rotors. This configuration of overlapping quadrotors is motivated by recent progress by a small company in the United Kingdom. The present work also has a number of interactions with researchers at AFDD and University of Maryland. A conference paper was written based on the initial stages of this research.

The Army's premier analysis tool, RCAS, is the platform being used to model the overlapping rotors. The thrust, power, torque, and required individual rotor rotation speed are being analyzed as the flight speed, turn rate, and center of gravity shift are varied. This project is focusing on the correct physics and applying rational methods. A conference paper was written based on the initial stages of the project, and the path forward seems appropriate. There may be opportunities to explore the vibration and load aspects of this unique configuration. The overlap of the rotors may cause impulsive changes in aerodynamic loading that lead to significant vibrations in forward flight. This possibility is something that the group could consider.

On-Demand Small Unmanned Aircraft Systems

The project title, "On-Demand Small Unmanned Aircraft Systems," is somewhat confusing. The project really ought to be described as the aerodynamic performance effort necessary to support a soldier tool box for providing a multiple number of combinations of quadrotor aircraft for field use. Specifically, most quadrotors are designed based on the concepts of fixed-pitch propellers. Because these rotors have fixed pitch, they generally incorporate variable rotations per minute to affect thrust change.

The project researcher was correct in noting that little codified data is available to describe propellers in edgewise flow. Although the University of Illinois has a reasonable data base for model propellers at the appropriate Reynolds number, these data simply cover the pure hover case. Consequently, the intent to develop small-scale, low-Reynolds-number, edgewise data at higher speed makes sense.

Although the use of simple blade-element codes might appear simplistic, the approach is probably adequate for the intended use. Once these predictions are reasonably validated during wind tunnel and limited flight tests, a family of quad raptor propulsors can be eventually finalized, printed, and incorporated into the soldier tool box. The appropriate mission-level vehicle capability software package (payload, endurance, speed, range) needs to be outlined in a timely manner.

Design and Characterization of Stretchable Electronic Materials and Components as Soft Robotics Enablers

This is a very ambitious project aimed at creating stretchable electronics using micron-scale particles embedded in soft elastomers to enable soft robotics. A range of particle aspect ratios (spheres to rods) is used to address various and interesting application spaces. The diagnostics developed in-house by the team is very unique and a definite plus for the program. Surface functionality and dispersion have not yet been considered, but their importance is recognized; studies of these effects are planned, in particular, as lower-length-scale particles will be examined.

By tuning the aspect ratio of the microfiber, the researchers have demonstrated the ability to achieve positive resistivity, negative resistivity, and also sharp resistivity changes with changes in strain. An impressive goal is the intention of demonstration of the ability to create a logic gate, which has interesting applications for simple, low-level control in soft robotic systems.

The researchers need to consider surface functionality and dispersion, as planned—which can go hand in hand with their present work—and will be key in order for the properties of these composites to be investigated. It might also be interesting to investigate the applicability of the group's work in this space with roll-to-roll printing using graphene for flexible electronics. The researchers need to address durability and longevity of the electronics performance over time and use. Also, the researchers need to thoroughly consider the literature in the flexible electronics space, which has grown rapidly in recent years.

Neuromorphic Control, Theory, and Hardware

This project investigates a novel approach for the design of feedback control circuits based on neural networks. The focus is on low size, weight, and power; biologically realistic neuron models; and low-complexity interconnections. Stable, closed-loop behavior is demonstrated for a simple, inverted pendulum. An ambitious objective is to understand low-level neuron circuit design principles. A concrete deliverable over the next 2 years is the design and fabrication of a memristor-based microcontroller and its testing in a closed-loop hardware setup.

It would be of great scientific interest if the investigators could develop a deeper understanding of circuit design principles. Some questions that could be asked are as follows: What is the closed-loop behavior of interconnected neurons? Can one predict the behavior of networks with an increasing number of devices? Scientifically, it would also be of interest to situate this novel approach within the broad field of neural network dynamics, such as the 2005 paper by Vogels, Rajan, and Abbott.[4]

Autonomous Navigation and Work in Three Dimensions

This work incorporates a system involving a high degree of freedom propulsion core (array of propellers), mechanical manipulator, and novel end effector. The latter utilizes a unique concept to vacuum grasp a selected object. The system is intended to be robust and capable of operation over a wide range of forces. The work concentrates on the effector (grasper), which utilizes a novel approach to assure flexible engagement and offers a robust capability to avoid inappropriate disengagement. This capability was demonstrated by experiment, which validated the design analysis. However, the work did not include adequate control discussion of the vehicles autonomous positioning within the control space. Structural dynamics ought to be considered for the truss. Incorporation of the important control/actuator bandwidth, proximity sensor, and system stability were not presented. Having said this, the effort looks very promising and is expected to incorporate a rigorous simulation, including the aircraft and robot manipulator, as soon as practicable. Development of a detailed stability model will also be

[4] T.P. Vogels, K. Rajan, and L.F. Abbott, Neural Network Dynamics, *Annual Review of Neuroscience* 28:357 -376, 2005.

extremely important and could include combined aircraft and robot control. Several current conference papers were cited that document the activity.

High-Fidelity Modeling of Dynamic Limbed Systems

This work is an innovative project focusing on design and controller development for legged locomotion mobility for robots. Legged robots could potentially enable faster locomotion, more agile movement, and the ability to handle disrupted environments. This project concentrates on leg design and the control, not on sensing or perception of environment, which is handled in other related projects. The test platform, entitled CANID, is a six-legged robot with a flexible spine. It has been demonstrated that the flexible spine leads to efficient robotic motion but also leads to complexity in control.

C-legs were implemented initially but also presented difficulties in modeling because they create uncertainty in terms of the unknown push-off point during motion. Thus, multiple other possibilities of leg geometries are being considered, with current focus on a five-bar leg with 2 degrees of freedom at the foot. Control algorithms have been developed—both with high-fidelity and lower-order modeling—the lower-order model will enable real-time control and is being validated against the high-fidelity model.

Many very interesting aspects of the legged robot are being considered, including how to design for both agility and speed, and a simultaneous ability to open a door, which are typically competitive goals. The ability to open a door adds weight in motors at each joint for dexterity, and that weight is counterproductive to attainment of speed. Control algorithms have been used in order to define the stable states in 9-dimensional space, and these are being used to enable the robot to move smoothly with changing gait and to address gaps in the landscape seamlessly. This is a very unique aspect of the work. The work, overall, is clearly defined, and this is an ambitious project with significant accomplishments and many areas of fundamental interest that remain to be explored.

Low Reynolds Number Dynamic Stall

With the Army's focus on small, autonomous aircraft in the future, the scientific community will need data on the dynamic stall of airfoils at low Reynolds numbers. Otherwise, it will be impossible to design these small aircraft, whether they will have flapping wings, rotating wings, or fixed wings. There presently are very little data in this area.

The classical National Advisory Committee for Aeronautics airfoil charts have a few plots down to a Reynolds number of 25,000; but these are for only a few airfoils, and there are less data at lower Reynolds numbers. In addition, the data in this area are only static data and would not indicate how these low Reynolds number airfoils would behave dynamically, which is essential for design of the aircraft the Army has in mind.

The researchers on this project have thought the issue through and have arrived at a research plan to obtain the data and also to obtain insight into the flows. The facilities that the researchers have developed will enable them to proceed along a well-designed path toward the needed database. The particle-image velocimetry data already obtained are very impressive. These data can give insight into when a vortex is shed and where it goes after it is shed. This type of information is what one will need to formulate lower-order dynamic stall models in this Reynolds-number regime.

Although the work so far is excellent, it will all go for naught unless actual load measurements are obtained. Partial loads can, in theory, be obtained from the particle-image velocimetry data. In particular, an integration of circulation and vorticity—around a closed loop that encircles the airfoil—ought to be able to give an approximation for the circulatory lift (although it cannot determine the complete lift, drag, and pitching moment). Nevertheless, the team needs to pursue this path, since it will give some preliminary data and will also give a benchmark against which to compare the load measurements that will follow.

The team's plan of attack on how to obtain loads is very sound. The concept of balancing a free-floating airfoil with counter-weights goes back to the Wright brothers and is a very powerful method. This method is very creative and will probably be able to extract out the steady lift, drag, and pitching moment. It remains to be seen if the method will be successful for obtaining the unsteady loads, but it seems that the seeds for success are there. The research team needs to proceed at full speed on this work.

Energy and Propulsion

Energy and propulsion focuses on the exploitation of innovations in energy sources, storage, generation, conversion, transmission, distribution, and management to provide technologies and configurations to improve the operational effectiveness and efficiency of Army platforms to enable military power projection superiority. The principal activity in energy and propulsion are specific research and development (R&D) programs in the areas of fuel injection and combustion, turbine engine efficiency and sand tolerance, lightweight hybrid gears, and ARL-developed power electronics components. KCI consists of research on advanced power electronics conversion and control. CCE includes advanced switching and control for power electronics and high-power-density and energy-efficient drivetrain technologies. The KCI is directed at advancing tactical energy networks (intelligent distribution and control and the integration of generation, renewables, storage, and distribution). The CCEs focus on high-performance packaging and thermal management and DOD-specific diagnostics for high-pressure and high-temperature combustion vessels and high-speed chemiluminescence.

Hybrid Gears

The overall objective of this research effort is to enable weight reduction of power transmissions, thereby increasing power density. In order to accomplish this objective, experimental investigation of hybrid gears under adverse conditions is being conducted. This research effort is of high relevance and has been ongoing for a few years. In addition to the potential weight reduction, hybrid gears may also provide vibration and noise reduction.

The experimental programs being conducted are of high quality. However, in absence of a companion modeling effort, these programs get reduced to conducting tests rather than experiments. For example, previous testing with subscale gears does not provide a high-confidence path for pre-test predictions for tests conducted with full-size gears. A modeling effort would help in the identification of relevant dimensionless parameters for the phenomena of interest. It would also provide insights into how to use experiments to understand fundamental phenomena and predict the behavior of full-scale transmissions. Sustained resources for the modeling portion are now being deployed.

Tribology and Lubrication Science

The research program on tribology is outstanding in its approach and content. This field has traditionally been very empirical in its approach. It has been common to gather data for material pairs of interest in the laboratory environment and then use the data in design of the components and systems. This approach can result in expensive problems because the laboratory testing may not encompass all the conditions for the system accurately. The research team has chosen to conduct experiments rather than tests. This approach is focused on getting answers to specific questions aimed at understanding the basic physics underlying the wear phenomena. This will make it possible to apply the knowledge to the conditions corresponding to the actual system operating conditions in the future. Experiments are logically planned with an analytical approach. Replacement of steel by lightweight Si_3N_4 could be beneficial.

The research program has two stated objectives: (1) achieve fundamental understanding of limiting failure mechanisms in power transfer interfaces to enable operation under adverse lubrication conditions and (2) increase power density and specific power in vehicle transmissions.

The research meets the first objective very well. Even though there is not a clear link to meeting the second objective, this is somewhat ground-breaking research. Its value will only be better understood once one has all the information and focus can be turned to utilizing it in the design process.

The hypothesis, as stated in the poster presented to the panel, is somewhat vague. In particular, the link between high-slide tribology experiments and specific gear or bearing applications needs improvement.

Plans to perform in situ diagnostics at high-sliding junctions are good. Application of more comprehensive experiments to characterize tribologically formed films would be good.

The team of researchers working on this program also demonstrate excellent effort in keeping abreast of the state of the art via attendance at conferences and other training opportunities. This approach needs to be continued to maintain relevance of the program and internal efforts. While the in-house activities do not have an extensive modeling component, there are excellent collaborative programs aimed at developing companion modeling capability. It will be very useful to include these collaborative activities in future reviews.

Effect of Semi-Molten Particulate on Tailored Thermal Barrier Coatings for Gas Turbine Engine

The overall objectives of this research effort is to enable uninhibited operation of gas turbine engines in harsh environments that may contain dust, sand, salt, ash, etc. The research effort is directed at understanding fundamental mechanisms of thermal barrier coating degradation caused by these contaminants. A balanced approach consisting of experiments and modeling (being pursued at partner institutions) is being executed. This problem is highly relevant to the gas turbine industry and goes well beyond just Army applications.

The experimental programs being conducted are of high quality. The experimental facilities are state of the art. Researchers need to pay close attention to ensuring that the configuration of the test coupons (geometry as well as film-cooling flow rates) are in the right parametric space and correctly simulate the fluid mechanics in gas turbine engine components. Another opportunity for improvement is to closely examine the concept of using average sand in the experiments. Such an approach may provide data that is not applicable to any specific conditions encountered by the engine. It may be a better idea to identify the range of parameters and conduct experiments that bracket the range.

Nonlinear Ultrasonics and Advanced Sensing Methods for High-Temperature Propulsion Materials

Overall, the quality of the work presented was excellent. The paramount goal of this work is to develop an in situ temperature-strain sensing capability at temperatures above 1250°C. Such conditions contribute to what is generally an austere sensing environment.

Ultrasonic frequency was selected to be sufficiently high so that the wavelength (speed of wave or frequency) is much smaller than the damage feature to be detected.

In this research, classic Fourier harmonic wave analysis concepts were used to examine a nonlinear quantity that is proportional to the ratio of the second harmonic wave amplitude and the first harmonic wave amplitude squared and also inversely proportional to the distance traveled by the wave.

This ratio is sensitive to microstructural damage accumulation. The researcher showed that this ratio increases with increasing fatigue cycles. It was also demonstrated that these ideas may be useful to detect thermomechanical fatigue damage precursors—for various materials (e.g., aluminum, titanium, and nickel based alloys).

The problem being addressed in this work—reliable damage detection for propulsion applications—is very important for current and future warfighting missions. The sensing methods under development can be used for damage detection, state assessment, and health monitoring. As such, they can eventually play a critical role in ensuring the safe operation of mission-critical propulsion systems.

Opportunities for this work include the following: (1) carefully comparing results with existing methods (e.g., eddy current and fluorescent penetrant inspection); (2) developing a benchmark (e.g., with a very precise type of damage) for rigorous comparison purposes—such a benchmark can be used to systematically compare alternative approaches; (3) examining relevant signal to noise ratio issues and possibilities for real-time/post processing; and (4) examining possibilities for use of such a sensor for future real-time control (e.g., reduce control system bandwidth when damage is detected).

JP-8 Desulfurization

In this work, the link to a pressing Army need is clear, and the approach is well chosen. The context to other non-ARL work on fuel desulfurization is good. The methodology is well chosen. The accomplishment of nearly 10 times improvement in organosulfur adsorption specific capacity since 2012 is commendable.

Opportunities for improvement exist in formulating the hypothesis, which is not well formed. The selection process for creating improved sorbents is ad hoc and could benefit from more rational design principles. A companion computational modeling effort could be beneficial here.

Transient Thermal Management of Electronic Components

Overall, the quality of the work was excellent. The paramount goal of this work is to better address traditional or classic thermal overdesign issues by using predictive knowledge of transients.

The work focused on examining fast transients (20 msec to 100 msec) for high-power applications. Multilevel encapsulated phase-change material (guided by classic low-pass filter modeling ideas) was used to appreciably dampen high-speed transients. Enhanced package design with greater than 40 percent peak temperature reduction was achieved for a benchmark case. The work was published, with one of the papers being highly cited.

The problem being addressed in this work, thermal management of electronic components, is very important for current and future warfighting missions. Thermal management methods can be used for controlling overall system performance, reducing design conservatism, and extending life. As such, this work can play a critical role for ensuring the safe operation of a mission-critical subsystem.

Opportunities for this work include the following: (1) gathering data from real devices and comparing it to simulation data, which can then be used to improve modeling capabilities; (2) partnering with Texas A&M University on the above and more precise modeling; and (3) showing how work can be used to better control and significantly impact the performance of an overall warfighting system or critical subsystem, which can include using feedback to precisely control transients in the presence of uncertainty.

Solid-State Circuit Breaker

Overall, the quality of the work was excellent. The paramount goal of this work is to design and build a solid-state circuit breaker with a much longer life expectancy than state-of-the-art nonsolid-state breakers while delivering improved performance. The work focused on addressing 850 V and 100 A breakers using silicon carbide. With a footprint of the same size as a mechanical breaker, the silicon carbide breaker achieved a factor of 6,500 improvement in switching time, while trading off resistance by a factor of 23.25, which is 3 times more efficient than silicon.

Avalanche diodes were used to accommodate 4 times greater surge vis-à-vis silicon. It needs to be emphasized that the developed solid-state breaker is not available commercially. The design procedure is fast—taking approximately 10 minutes for a redesign—and the desired switching profile can be programed in, for example. The research results have been published.

The problem being addressed in this work—a longer lasting circuit breaker—is very important for current and future warfighting missions. A longer-lasting circuit breaker can be very useful for ensuring and controlling overall system performance and extending system life. As such, this work can play a critical role for ensuring the safe operation of mission-critical systems.

Opportunities for this research include the following: (1) carefully quantify performance vis-à-vis state of the art; (2) the possibility of developing a graphical user interface to permit users to input desired characteristics; (3) working toward automating the building process; and (4) showing how this breaker (or the next version of the breaker) can make a significant difference vis-à-vis other competing breakers within a critical warfighting system application.

Advanced Thermal and High-Voltage Electronics Packaging

Overall, the quality of the work was excellent. Current power electronics packages are generally oversized and have multicomponent reliability challenges. A critical goal here is achieving a greater power density. The state-of-the-art package is one with the following specifications: 150 W/(sq cm) power density, 10 kV operational voltage, and 150°C operational temperature. A 2-year goal for the research is as follows: 1 kW/(sq cm) power density, 30 kV operational voltage, and 200°C operational temperature.

An integrated power tower package has been developed—a new packaging approach that stacks devices between multipurpose mechanical/electrical/thermal copper interconnects. While commercial packages have a characteristic thermal resistance of 1 to 3 (°C/(W/(sq cm))), the researcher has achieved 0.2 and is aiming for 0.1. Modeling has shown that this goal requires that multiple methods be properly combined.

The researcher has shown that one can get close with a heat spreader (i.e., thermal ground plane). However, more is needed in order to achieve the objective. The researcher worked with WMRD to develop a substrate board using custom cold spray additive manufacturing powder to replace more standard direct-bond copper board that cannot operate at high voltages. The research achieved 500 W/(sq cm) and 30 kV with the above additive manufacturing approach. The research results have been published and have received best paper track and session awards.

The problem being addressed in this work—the design of thermally robust electronic packages—is very important for future warfighting missions. A more robust electronic package can be very useful for ensuring and controlling overall system performance and extending system life. As such, this work can play a critical role for ensuring the safe operation of mission-critical systems.

Opportunities for this research include the following: (1) pursuing performance-based modeling (i.e., co-design co-simulation work) with faculty from the Naval Academy in order to more precisely and holistically optimize thermal, electrical, and mechanical design objectives; (2) collaborating among others with University of Maryland, Power America, NASA, and the Department of Energy on the above; and (3) establishing future longer-term benchmarks and goals.

Spray Combustion Studies

ARL has a commendable program on spray atomization and combustion. It covers both computational and experimental topics and is staffed by several competent early career researchers. The facilities are impressive. The program involves the investigation of diesel-engine spray combustion by studying pulsed, round, fuel jet injection into gas chambers—at pressures ranging up to 60 bar at this point. The combined computational and experimental effort aims at understanding spray dynamics, mostly in the context of combustion for propulsion. The scientific questions being addressed are well formed, and the findings to date were clearly stated and shown in the posters presented to the panel. The results are relevant to combustion science, with the computational and experimental collaborators on a clear path to generating useful insights. The team appears to be well connected with multiple consortia both within and outside DOD, both nationally and internationally, focusing on the study of fundamental combustion of diesel sprays. The team is to be commended for the tight integration of theory, computations, and experiments.

The team's three-dimensional, unsteady computational model currently addresses the lower pressure conditions by considering both the gas and liquid to be incompressible and isothermal (with no coupling to energetics). Vaporization was not considered. The surface was tracked by the volume-of-fluid technique. Examples of their progress are given by several papers that include the results of collaborative research. See, for example, Ma et al. (2014),[5] Bravo et al. (2016),[6] and Bravo et al. (2016).[7] The round jet was previously studied by Shinjo and Umemura (2010)[8] for spatially developing instability and by Jarrahbashi and Sirignano (2014)[9] and Jarrahbashi et al. (2016)[10] for temporally developing instability. The ARL study can build upon these works in several ways. For example, a wider range of fuels and, therefore, of Ohnesorge numbers can be studied, and stronger coupling with experiments can be made.

---

[5] P.C. Ma, L. Bravo, and M. Ihme, "Supercritical and Transcritical Real-Fluid Mixing in Diesel Engine Applications," pp. 99-108 in *Center for Turbulence Research Proceedings of the Summer Program 2014*, https://web.stanford.edu/group/ctr/Summer/SP14/05_Two-phase_flows/06_ma.pdf.

[6] L. Bravo, S. Wijeyakulasuriya, E. Pomraning, P.K. Senecal, and C.B. Kweon, Large eddy simulation of high Reynolds number nonreacting and reacting JP-8 sprays in a constant pressure flow vessel with a detailed chemistry approach, *Journal of Energy Resources Technology* 138:032207-1, 2016.

[7] L. Bravo, D. Kim, F. Ham, K.E. Matusik, D.J. Duke, A.L. Kastengren, A.B. Swantek, and C.F. Powell, "Numerical Investigation of Liquid Jet Breakup and Droplet Statistics with Comparison to X-ray Radiography," AIAA Preprint, 52nd AIAA/SAE/ASEE Joint Propulsion Conference, http://arc.aiaa.org/doi/abs/10.2514/6.2016-5096, 2016.

[8] J. Shinjo and A. Umemura, Simulation of liquid jet primary break-up: dynamics of ligament and droplet formation, *International Journal of Multiphase Flow* 36:513-532, 2010.

[9] D. Jarrahbashi and W.A. Sirignano, Vorticity dynamics for transient high-pressure liquid injection, invited paper, *Physics of Fluids* 26(10):101304, 2014.

[10] D. Jarrahbashi, P.P. Pavel, W.A. Sirignano, and F. Hussain, Early spray development at high-density: Hole, ligament, and bridge formations, *Journal of Fluid Mechanics* 792:188-231, 2016.

The work appears to mostly focus on spray expansion and ignition in large chambers without considering the dynamical changes that would occur in an operating piston engine. In this regard, the researchers' plans to extend their experimental diagnostics to include an observation port on an operating piston engine are well thought out and sound.

The team needs to seek different domains of physical behavior in the plot of Weber number versus Reynolds number. What is the path for the cascade through decreasing scales as the two-dimensional Kelvin-Helmholtz (K-H) waves become three-dimensional with eventual droplet formation? The various, smaller three-dimensional structures consisting of lobes, holes, bridges, ligaments, and droplets could be identified. Specifically, the team could study how the Ohnesorge number relates to the break-up mechanism.

The team needs to do post-processing to determine the vorticity field and to relate the vortex dynamics to the development of the liquid-gas interface dynamics, which could explain the physical mechanism for the break-up cascade. Then, post-processing analysis could be used to determine the major vorticity generation terms, especially for the streamwise vorticity and the hairpin vortices formations.

The team plans to develop, in the near future, a computational model that describes the injection of fuel at subcritical temperature into a gas at supercritical pressure. The team understands that this will require one to consider the use of a cubic equation of state, variable density, variable composition, heat and mass exchange between the two phases, an enthalpy departure function to correct for real-gas behavior, the energy of vaporization (which is distinct from the latent heat), and detailed phase equilibrium to account for both the original gas in solution and the fuel vapors in the gas. A review of the work done at the Air Force Research Laboratories (Edwards Air Force Base) by Talley, Chehroudi, and co-workers would greatly benefit this effort.

An issue for discussion and examination is the need to use large-eddy simulations (LES) for the liquid jet problem. The transitional turbulence generated by the liquid jet has two major, critical features. (1) Transitional turbulence is the cause of the hydrodynamic instability and leads to jet break-up process; therefore, it cannot be treated as a subgrid phenomenon. The team clearly understands this point. (2) The turbulence kinetic energy generated upstream in the orifice in the absence of cavitation has substantially smaller magnitude. Thus, the value of modeling the smaller scales is questioned. Note that the three papers from non-ARL work referenced above use direct numerical simulation without resolving turbulence generated in the orifice flow. The ARL team is not modeling the ambient turbulence that might be generated through air intake. However, if it did, the length scale would be larger than the liquid jet dimension—implying that LES will not help there.

The experimental work addresses interesting issues concerning the transient injection of fuels with vaporization, mixing, and chemical reactions. Higher pressure tests are included. While excellent high-pressure facilities are available at ARL, a potentially important collaboration with Argonne National Laboratory has been made to access testing with higher-resolution X-ray imaging, albeit limited to atmospheric conditions. The fuels include JP-8, diesel oil, and biofuels. Simultaneous diagnostics at high speeds are available. Ignition and ensuing, transient, combustion are studied. Comparisons with computations are made yielding qualitative agreement but with quantitative disagreement. The researchers agree that pursuit of higher resolution is critical. Investments have been made in forefront imaging to accomplish this goal.

Several improvements to the research can be made. The team could do the following:

1. Determine the three-dimensional, smaller-scale structures that form in the transition from two-dimensional (2D) K-H waves to the droplets;

2. Examine whether the path from 2D K-H waves to droplets depends on Ohnesorge number rather than just on Reynolds number and Weber number;
3. Attempt to identify whether fuel-rich "blobs"[11] exist and, if so, can a pattern be found related to the original K-H wavelength;
4. Determine if vortex structures are forming in the gas and study the role they play in the mixing and combustion processes (velocity measurements will be helpful here);
5. Use simple injector configurations (e.g., simple orifices) in the initial studies instead of commercial injectors to allow investigation of injector design effects;
6. Invest available resources to study of the atomization, mixing, and the locations within and outside the sprays where local ignition occurs, although it is doubtful that the proposed detailed chemical kinetics studies would impact the results, considering the complexity of fuel sprays combustion; and
7. Obtain more information on how high-speed chemiluminescence would yield information about ignition and the quality of the combustion process.

Logistics and Sustainability

The logistics and sustainability area focuses on empowering the Army with breakthrough technologies and capabilities to conduct expeditionary maneuvers with substantial reduction in operation and sustainment costs. Its goal is reduced-maintenance and longer fatigue life Army systems. Condition-based maintenance is driven by research in (1) material state awareness/self-healing, (2) digital nanomaterial architecture (additive manufacturing), and (3) material damage precursor development (failure correlation) forming the virtual risk-informed agile maneuver sustainment (VRAMS) program. This drive advances in condition-based maintenance to mission-informed material state-based awareness. This represents an advance from hardware-level concern to a focus on higher-level, mission-relevant operations. VRAMS has developed a 10-year plan that includes demonstration and transition to technology readiness level (TRL) 5. The 5-year plan is supported by uniquely integrated computational, modeling, test, and analysis facilities and environments.

The main research activities of the logistics and sustainability group are centered around the VRAMS program. VRAMS emphasizes early damage detection, focuses real-time integrity monitoring and state awareness, expedites operation adjustment, and automates planning and implementation of needed logistical modification. This is a laudable and very desirable goal for the Army. ARL is to be commended for tackling this holy grail of integrated material system behavior.

VRAMS is a paradigm change in design, operation, and sustainment for all structural platforms. The program has a grand and ambitious vision with huge industry-wide impact. It involves multidisciplinary research with a link to current emerging research topics on information, sensors, multifunctional materials, intelligent structures, multiscale, multi-physics computations, etc. It touches multiple research areas with vertical integration possibility.

Challenges for the logistics and sustainability area include the following: (1) Enormous efforts are required with focused steps to execute the program. Phase in approach is needed with clear near-term and long-term objectives and targeted milestones or metrics. (2) Because of the complexity and diversity of the program, integrated efforts in both technology and industry-academia collaboration are needed.

While the program has laudable high-level goals, its extreme breadth can also make progress difficult. ARL has reached out to collaborate with several universities and the Navy and could benefit greatly

[11] A "blob" is some combination of liquid and fuel vapor.

by expanding that outreach to many other leaders in numerous programs with similar goals, including industry and the Air Force. Having identified that major challenge, a next logical step for ARL might be to harness the collective wisdom of the broader community as to proposed strategies for approaching such a major goal by asking one or more professional societies to convene a major conference and harnessing the brainpower of all sectors—industry, academia, and government—for such an undertaking.

Sustainable, long-term support would be critical to carry out this research. Strong interaction with Army Research Office program managers are needed to develop collaborative programs. For instance, VRAMS-focused programs as part of multidisciplinary university research initiatives might be sought to create additional infusion to the project. Open-campus programs can be utilized to address project needs. Joint efforts with academia and industries need to be further encouraged. A benchmark test-bed approach is needed to demonstrate technology development to align with DOD-targeted thrust areas. ARL, in the short term, needs to focus on those known parameters for a proof of concept, while continuing long-term studies on unknown parameters, such as damage precursors on selected materials (metals and composites) that are most critical to the rotorcraft industry. For instance, as an intermediate point, some explicit linkages need to be made between damage precursors and sensors, for specific materials, and damage mechanisms and goals need to be set in that context. Early interaction with airworthiness authorities is needed. Promotion of VRAMS concepts and results is needed at conferences, in journal publications, and at special VRAMS-focused technical sessions and meetings.

The approach initiated by ARL has several laudable elements, as discussed below. All projects need to define intermediate goals where specific materials (or classes) are identified and specific failure mechanisms are linked to clear structural changes. These structural changes then need to be linked with sensing and/or health monitoring technologies that can sense specific damage precursors. These mechanistic connections between explicit structural changes and failure mechanisms and sensing strategies can be strengthened significantly.

Materials Damage Precursor

The concept of very early detection of structural material damage to enable early action for safety and maintenance is a very desirable goal. Because it is so strongly related to material microstructure, participation by materials scientists in this program is essential. It is not clear that testing individual fibers adds significant insight into failure modes in fiber composites, nor is it clear what precursor structural changes are being tracked, how this can be sensed in a fielded application, or how these will relate to eventual failure. Further, phenomena occurring on the surface of a fiber is not indicative of fiber-matrix interfacial failure modes. Clear understandings of the envisioned failure mechanisms relating this to measurements would accelerate useful progress.

Elastodynamic Modeling of Metal Microstructure

The program approach is relating material behavior to microstructure, which is an important subset of this effort. There is significant knowledge both from theoretical work and experiments as to how stress affects the elastic stiffness tensor of crystals, and several models can be used to extend this to the behavior of polycrystalline aggregates. The program would benefit from more clearly defining how this work is differentiated from work now in the literature and presenting an explicit sensing strategy by which this information could be used in measuring stress in fielded applications. The effort would also benefit by including materials engineering talent and evaluating the extensive literature in that area.

Advanced Sensor Fusion

This program is a well-thought-out analytical integration of measurable and detectable damage and structural behavior. It provides a very tangible example of very advanced and powerful use of sensors with a clear path for in situ, in-mission use. The collaborations with airframe manufacturers and universities is particularly laudable. The research has captured an important challenge and has made a strong contribution.

Magnetostrictive Material Modeling for Structural Health Modeling

This project presents a novel method for local measurement of stress using the magnetic properties of a Fe-Ga alloy. The work is well thought out and uses an advanced process—magnetic pulse welding—to join the alloy to the prototype component without the use of heat. The program is clearly carried out carefully, intelligently, and with academic impact. It is not clear, however, what the strategy might be to incorporate this technology into fielded systems. Addressing this at a conceptual level would better tie this program to the broad VRAMS mission.

Reduced Logistical Burden by Part Reduction via Composite Drive Shafts

This program proposes the possibility of reducing weight and part count by replacing metal drive shafts. These shafts require flexible couplings with longer composite shafts having an elastomeric matrix to accommodate the motion carried by the couplings. The replacement of aluminum with composites can provide significant rotorcraft weight reduction, making it more capable. Because the primary goal of eliminating couplings requires demonstration of functionality of the composite shaft, it is not clear why the first part of the program deals with ballistic vulnerability and puts off the demonstration of functionality. This work can provide real benefits to the warfighter. Considering full drive shaft designs early in the design of the program can accelerate impact.

Topology Optimization for Additive Manufacturing

The researcher that presented this topic to the panel is commended for a very clear and insightful articulation of the work. The development of topology optimization is a very worthwhile and relevant effort and does lend itself to additive manufacturing. It was not fully clear what the novel aspects of this work are in the rapidly developing areas of topology optimization and additive manufacturing. This work would benefit by developing an extensive literature review in these areas. In addition, including additive manufacturing and materials expertise in this project would benefit the development of relationships among process, structure, property, and performance that are required to ultimately apply topology optimization through additive manufacturing in fielded applications.

Damage Precursor Detection and Identification in Composite Materials

The goal of linking damage propagation prediction with actual data is a very desirable goal and a challenge in composite materials. It is very well known that, in composite systems, compliance will increase with damage. It is not clear how this work adds clearly to that body of knowledge. Because it is so strongly related to material microstructure, participation by materials scientists in this program is essential.

OVERALL QUALITY OF THE WORK

Vehicle Intelligence

In each of the three pillars of the vehicle intelligence (VI) program—intelligence and control, perception, and human–robot interaction—the research quality was generally high. Research results are published in high-quality journals. Collaboration with other government agencies, industry, and universities continues to yield benefits. Internal personnel advancement, including hiring new, well-qualified Ph.D. researchers, strengthens the capability of the Sciences for Maneuver VI R&D program.

Each of the three pillars of the VI program has demonstrated significant progress in advancing its R&D objectives to support the warfighter in increasingly complex environments. The R&D activities in each pillar were consistent with its defined objectives. Opportunities in multiperson/multirobot scenario simulation, teaming of autonomous systems with soldiers in uncertain environments, multispectral sensing, range sensing, contact sensing, and immersive display of robot lidar imagery may allow ARL to take the lead in this research and offer greater benefit to the soldier.

The intelligence and control pillar employs innovative approaches in developing and supporting advanced technologies, algorithms, and tools in support of the warfighter effort. There are high-value collaborations with top universities, industry, and other government agencies. This pillar invests in advancing the effectiveness and efficiency of its research personnel.

The perception pillar benefits from high-quality collaborations with top universities that enable successful hiring of outstanding personnel at the Ph.D. level. Current trends and research vectors were observed. Research personnel participate in highly competitive technical conferences and publish in top research journals.

The human–robot interaction pillar's R&D program is of high quality and based on rigorous design and appropriate metrics. It benefits from a substantial increase in external and internal collaborations. Through early retirements and expansion of the postdoctoral program, qualified Ph.D. personnel have been hired.

The VI R&D program is correctly constituted and resourced with the ARL workforce and facilities. In general, the VI team demonstrates good awareness of the scope and direction of R&D in each of the pillars. Cognizance of related activities in industry, government, and international R&D enable meaningful goal setting and tactical adjustments in specific program advancements.

Within VI R&D programs, research quality is generally of high quality. Based on rigorous design, useful evaluation and analysis, appropriate metrics, thorough understanding of related research, workforce interaction at critical open conferences, and publishing VI investigations in top journals, VI is well positioned to maintain and improve the quality of its research products.

Recently hired researchers within VI appear to be well qualified to conduct leading R&D in VI. These new additions to the VI workforce have been educated and trained by leading faculty in the three pillars at top-ranked U.S. academic institutions. At ARL, new personnel are exposed to effective mentoring. The VI principal investigators are well prepared and energetic.

VI collaborations with U.S. industry, government, and academia appear to be extensive and are very effective at advancing the VI R&D mission. These collaborations are important components driving VI awareness, leading to the establishment of meaningful goal setting and tactical program adjustments. Similarly, the collaborations feed the energy of the VI principal investigators.

Inclusion of U.S. soldiers in VI field experiments is commendable. Usage of more realistic vignettes and real-life simulations in experiments would be very beneficial. In particular, the use of realistic warfighting vignettes, where researchers are in the field with soldiers, provides opportunities to test and evaluate research hypotheses more thoroughly, including the revelation of previous unknowns.

Within VI, the emerging shift from one-person/one-robot studies to multiperson/multirobot studies merits sustained attention. This shift exposes VI to more complex teaming architectures, a concomitant realistic field environment, and potential improvement of the validity and applicability of the research results.

Some strategic goals and tactical milestones for VI R&D programs could be made more apparent. To help quantify general progress and application-specific performance, more efforts need to be made in terms of baselining and benchmarking. A process whereby desired capabilities and goals are broken down into a sequence of achievable (realistic) short-term capabilities and goals would be beneficial. Unifying demonstrations, milestones, objectives, and capabilities could help to better motivate the specifics being developed and how they will be integrated.

Application of more systems integration principles across research projects and pillars would strengthen the overall impact of VI research products. Similarly, connectivity between individual principal investigators could be improved.

The VI program is well positioned to maintain and improve the quality of its research products. To move to the next level, VI needs to undertake carefully chosen, audacious, grand challenges that go beyond the extant state of the art. Resulting activity and research products would provide leadership in R&D. This would yield inherent advantages in framing VI problems to achieve solutions that benefit the Army.

Vehicle Technologies

In each of the foundational pillars and key enablers of the vehicle technology program—platform intelligence, energy and propulsion, platform mechanics, and logistics and intelligence—the research quality was generally high. Research results are published in high-quality journals. Collaboration with other government agencies, industry, and universities continues to yield benefits. Internal personnel advancement, including hiring new, well-qualified Ph.D. researchers, strengthens the capability of the sciences-for-maneuver R&D program.

In the platform mechanics area, the research was overall of an excellent quality. The steady improvement of the work over past years is quite evident and is to be commended. Each of the tasks was well thought out. The tasks have an appropriate amount of connectivity with other tasks in this area, and the researchers have a good understanding of how their work fits into the larger picture. Although the publications of the researchers at conferences and in journals continue to improve—as they have over the past few years—this area can still use more improvement. Hopefully, more travel funds and better mentoring will see an increased number of publications. Overall, the work is excellent and shows a solid improvement in the research quality, applicability, and dissemination of the research results.

The Koopman decomposition of periodically excited Hopf bifurcation research (nonlinear system theory) initiative is outstanding, with a potentially significant impact on Army understanding and exploitation of nonlinear mechanics, such as dynamic stall, Floquet instabilities, low-order modeling of control system design, and ground resonance instabilities.

The experimental and analytical research on continuous trailing-edge flap actuation authority is important, well executed, and advances the state of the art in understanding the rotor-to-rotor interactions and wake-to-wake interactions. The sound methodology of simulation, analysis, and experimental testing is to be commended.

The R&D work on stretchable electronic materials is outstanding. This work is a clear demonstration of ARL in-house research leadership. The demonstration of tuning the aspect ratio of microfibers is an advance in capability to achieve positive, negative, and also sharp resistivity changes with strain.

In general, platform mechanics research presentations were professional, logical, content-rich, and useful. Growth in knowledge content by researchers and staff is noted. Significant advances in the use of experimental and diagnostic facilities were observed (high-altitude test chamber, tribology upgrade, low-turbulence wind tunnel).

In the energy and propulsion area, clear advancement has occurred since the panel visit 2 years ago. The new facilities are producing publishable results. The early career researchers show greater maturity, knowledge, and subject command. In the high-pressure spray combustion domain, ARL is well positioned for a leadership role. Wider recognition of this group is needed over the coming years.

The ongoing hybrid gears research is excellent. This work has high potential to reduce system weight, vibration, and gear noise. A modeling effort that includes the development of primary dimensionless parameters is needed.

The research on the effect of semi-molten particulate on tailored thermal barrier coatings for gas engines is also to be commended. Progress obtained in collaboration with industry is advancing the understanding of thermal barrier coating degradation caused by dust, sand, salts, and ash.

Research and development of nonlinear ultrasonic and advanced sensing methods for high-temperature propulsion is excellent. With a goal of an in situ temperature-strain sensing capability at temperatures above 1250°C, the proposed ultrasonic frequency method is quite appropriate and useful.

The overall quality of the transient thermal management of electronic components is excellent. This work is critical to warfighting missions. Using multilevel encapsulated phase change material, this work focuses on examining fast transients for high-power applications.

A big picture demonstration showing how the energy and power work will be integrated would be helpful to the researchers and for future reviews. Such a demonstration can be used to pursue a very relevant and informative system-of-systems analysis that can be used to identify weak links, guide research, and steer collaborations. This team needs to discuss the limitations of approaches being taken in each of the research initiatives. Such a discussion will fundamentally enhance understanding for early career researchers.

In the logistics and sustainability area, while the program has laudable high-level goals, its extreme breadth can also make progress difficult. ARL has reached out to collaborate with several universities and the Navy and could benefit greatly by expanding that outreach to many other thought leaders in numerous programs with similar goals including industry and the Air Force. Having identified that major challenge, a next logical step for ARL might be to harness the collective wisdom of the broader community as to proposed strategies for approaching such a major goal by asking one or more professional societies to convene a major conference and harnessing the brainpower of all sectors—industry, academia, and government—for such an undertaking.

Several logistics and sustainability research programs are laudable—namely, materials damage precursor, elastodynamic modeling of metal microstructure, advanced sensor fusion, and topology optimization for additive manufacturing.

7

Human Sciences

The Panel on Human Factors Sciences at the Army Research Laboratory (ARL) conducted its review of selected research and development (R&D) projects of the ARL Human Sciences Campaign and the Analysis and Assessment Campaign at the Aberdeen Proving Ground, Maryland, on July 14-16, 2015, and at the SFC Paul Ray Smith Simulation and Training Technology Center, Orlando, Florida, on June 14-16, 2016. The human sciences project areas reviewed were as follows:

- *Humans in multiagent systems*. These efforts aim to achieve critical technological breakthroughs needed for future Army multiagent, mixed-agent teams to effectively merge human and agent capabilities for collaborative decision-making and enhanced team performance in dynamic, complex environments. The challenges for human sciences R&D are soldier workload, situation awareness, trust, influence, and cultural cognition.
- *Real-world behavior.* The objectives of the R&D in this area are to enable the collection, analysis, and interpretation of human behavioral data within dynamic, complex, natural environments. ARL conducts R&D in the following two areas: (1) real-world complexity in human science experimentation and (2) assessing human behavior in the real world. A key focus of this work is the development of novel technology and methodologies and to collect and analyze these data in real-world conditions.
- *Toward human variability.* The goals of this R&D are to enable high-resolution, moment-to-moment predictions of an individual soldier's internal and external behavior and performance and the ways in which soldiers interact dynamically in mixed-agent team and social settings in both training and operational environments. Human variability R&D is conducted in the following two areas: (1) multifaceted soldier characterization to develop a comprehensive understanding of the factors influencing human variability and (2) brain structure function coupling

to create a multiscale understanding of the relationship among the brain's physical structure, its dynamic neurophysiological functioning, and human behavior.

- *Training*. The goal of the training program is to discover and develop methods, models, tools, and technologies that will increase soldier readiness by improving training methods and training technologies.
- *Integration technologies*. The objectives of the integration technology R&D program are to discover and innovate principles and mechanisms for the integration of humans and systems.
- *Augmentation*. The goals of this R&D program are to develop and enable technological approaches for augmenting fundamental human capabilities that may enhance Army mission-related performance.

The panel also reviewed in 2015 a component of the Analysis and Assessment Campaign portfolio on assessing mission capabilities of systems. The goal of this area, more or less, is engineering and acquisition decision support for current and future Army human systems. Efforts are conducted in two areas, the first of which is human factors, which focuses on integrated human factors engineering (HFE) and system engineering (SE) assessments and analytic techniques to predict human, system, and mission capabilities early in the acquisition cycle. In this first area, HFE applications and tools are developed and refined to lower acquisition costs and improve design. The second area, soldier survivability, is concerned with analysis and engineering to increase the survivability of platforms and soldiers operating in combat environments.

Just prior to the 2015 assessment by the Army Research Laboratory Technical Assessment Board (ARLTAB), ARL reformulated its R&D programs to align with its science and technology (S&T) campaign plans for 2015-2035. This transformation was in its early stages during the 2015 review but appears to have crystalized for the 2016 assessment.

The increased emphasis on understanding and managing real-world complexity within the key campaign initiative (KCIs) is an important and notable development. Increasingly, complex sociotechnical systems that encompass interactions among multiple agents whose behavior is governed by intangible variables (e.g., emotion and culture) will pose many tough challenges for the Army after next. Research in this area will likely promote increased interdisciplinary collaboration with other campaign areas. Conceptual and theoretical breakthroughs from the growing adaptive complex systems engineering literature need to be leveraged to support this work.

ARL responsiveness to prior recommendations from the ARLTAB was fully evident in both of these reviews. For example, the preparatory materials provided in advance of the 2016 review meeting were exceptionally clear, concise, and focused; presentations were generally more uniform and at a higher standard; research hypotheses were clearly stated, and experimental variables were described and discussed in detail; and access by researchers to subject populations with face validity has clearly improved but still remains an issue.

Proactive research management approaches were evident, such as the Human Sciences 6.1 refresh, which has led to a more cohesive program and the Big Idea process for stimulating and supporting innovative leaps forward in research.

In general, gains were evident in publication rates and in the establishment of collaborations and partnerships with relevant peers inside and outside ARL. The ARL human sciences work environments are exceptional in terms of their unique and advanced technological capabilities to support research. ARL has continued to successfully attract clever postdoctoral researchers from a diverse set of universities and disciplines.

Kudos are due the ARL leadership for their success at overcoming government-wide constraints on conference participation by scientists and engineers. This year, presentations at technical and professional meetings increased, coupled with substantial gains in published papers in the peer-reviewed literature. The proactive efforts to enhance leadership in the broader scientific community through data and resource sharing and professional engagement (e.g., editorships and participation in conference organizing committees) is a vital step forward in building long-term credibility and influence. Taken together, these are outstanding accomplishments and mark a visible advance over prior years.

ARL's vision of the future challenges to be faced by the Army is compelling and provides the foundational basis for investments in innovative R&D. Success in the execution of quality research that may deliver beneficial options to the Army after next faces significant challenges and opportunities. To overcome anticipated challenges and to leverage opportunities, ARL has developed and implemented an impressive array of organizational visions, initiatives, and strategic plans aimed at enabling environments and processes that leverage staff competencies and motivate innovation; encourage mentoring, coaching, and career management; and facilitate opportunities for collaboration with external peers (e.g., through the ARL open campus initiative).

While there was substantive evidence of good progress by ARL in all these areas, there were some areas that may require attention. Opportunities for technical and career mentoring do not appear uniform or consistent across the Aberdeen and Orlando sites. Orlando personnel need better connectivity with Aberdeen (e.g., Y: drive accessibility). The Big Idea initiative is a very good idea, but there was some evidence that bench-level staff in Orlando were not fully cognizant of the opportunities offered. While considerable progress has been made with increased publication in peer-reviewed literature, ARL staff expressed concerns that the public release clearance process is unduly bottlenecked, adding unnecessary delays and overhead in getting research to press. There appears to be a need for systematic feedback mechanisms from stakeholders—principally users/customers—as a basis for refining research and planning.

There appears to be a need for improved project-level planning that systematically considers Army requirements in the context of what others have done, are doing, and where ARL could best contribute. Good project planning that guides what problems researchers select to work on, how the work is executed, and the responsibilities assigned for transitioning the products of that work forward into the value stream is consistent with the management and execution of quality science and innovation. Available project plans would provide the ARLTAB assessment with valuable contextual data regarding research baselines, related advances, and expected unique contributions of the work under way and planned. A well-informed ARLTAB can better assess the quality of the science programs and address, for example, whether experimental designs and analyses are reasonable and appropriate to achieve desired technical objectives. Unfortunately, although project-level plans may exist, they were not provided to the ARLTAB.

A general challenge, noted in past assessments by the ARLTAB, persists with respect to the critical need for access to military-relevant subjects in the work. Many of the real-world research questions that ARL is dealing with urgently require more effort with respect to subject populations and the representation of mission contexts. A number of studies presented drew upon ARL researchers as subjects to a worrisome degree. As another example, researchers at the soldier performance and equipment advanced research (SPEAR) facility waited 8 months to acquire and run a limited number of soldiers as subjects. ARL needs to find a workable solution to this problem that threatens to compromise the credibility and impact of important research addressing vital Army needs.

On a positive note, statistical analysis at ARL has become increasingly rigorous; however, other analytical tools (e.g., mathematical modeling [optimization] and data mining) need further consideration in the program of work.

Several questions arose with respect to gaps in the disciplinary composition of the ARL human sciences workforce. Important core competencies, such as systems and simulation engineering, are absent; hiring is needed to grow these competency areas. There was also an expressed challenge and concern that the scant number of technicians that support the increasingly high-tech R&D within the Human Sciences Campaign is too low.

HUMANS IN MULTIAGENT SYSTEMS

Accomplishments

ARL human sciences work in multiagent systems addresses sociotechnical network operations, focusing on distributed collaboration and decision making in complex operational environments; sociocultural competency skills needed to support, inform, and influence operations in complex environments; human–robot trust; teaming of humans and intelligent systems; and human control of multiple robots. ARL appears primed to make important contributions to this paradigm and demonstrated a sound awareness of the key trends driving the research challenges in this area (i.e., the rise of the networked organization, increasing autonomy, and the need for cultural competency in military operations). The work in the area of human–robotic interactions and trust builds on the extant baseline of substantive theoretical and empirical work dealing with trust in automation. The conceptual organization of data-to-decisions at social cognitive, information, and communication layers allows for effective linkages and human–system integration with other multiagent systems activities. Hence ARL is applying a well-integrated systems approach to the study of purposeful social systems, addressing not only structure but also the role and function of sociotechnical Army systems involving people, information, and technology.

Challenges and Opportunities

The potentially valuable work in this area faces inherent challenges, but overall there does not appear to be a coherent vision for how the research holds together and cumulatively builds to push the state of the art. This may be the result of a forced merger of ongoing activities into the new campaign organization, which is still somewhat immature.

The problem area as scoped is very broad-ranging, including human–technology interactions to human–human interaction (sociocultural interactions), and it is not clear how all of the pieces fit together. For example, the distinction between the human–robot interaction (HRI) research and sociocultural research is an important one, because these groups have different pre-theoretical ideas and histories, concepts, analogies, and terminology and very different subject matter and methodologies. Although there are connections that could be made between these areas, they are generally distinct subdisciplines, and the connection between them needs to be made clear in the present organization.

The emerging work on joint operations of robots and humans is an important area of work with a large body of extant research that needs to be strategically considered as ARL evolves its niche in this area. There are many directions to be considered, including multiple HRI configurations (i.e., multiple humans with multiple robots), threats to security or integrity of robots as potentially compromised agents, and the possibility of coadaptation and coagency, where either the human or agent can take control. The progress at ARL has been unidirectional (i.e., human to robots) and needs to progress to encompass bidirectional communication—that is, humans trusting robots and robots trusting humans.

Good use is made of modeling and simulation (M&S) in this area, although there is room for some improvement by expanding the toolkit of available techniques. For example, there are instances in which a statistical or event-driven simulation is used instead of physics-based M&S. M&S has emerged as a discipline in itself, as opposed to only an enabling technology, with several universities now offering degree programs. As such, ARL needs to consider recruiting more simulationists onto its research staff.

The emphasis on experimental work by the human factors group is commendable, but good mathematical and computational models of the robots, the interacting humans, and the feedback could save a lot of trial and error, provide insight, allow testing of different control regimes and feedback strategies, quantify sensitivity, help with identifying worst-case scenarios and sources of instability, and provide guidance to the design of experiments. There are modeling opportunities, especially of human operators, of robots, and of human collaborators from the viewpoint of the robot. The development of such models can aid understanding about how humans and robots interact by using inexpensive simulations and sensitivity studies. These studies need to be done before undertaking costly experimentation with humans and physical machines.

The research staff, including postdoctoral researchers and interns with recent experience, appears competent. However, from a staffing perspective, this is an area that would benefit from growing interdisciplinary collaborations as the portfolio matures. Similarly, the modeling and simulation capability to support this work is good, but an expanded toolkit will be needed to enable substantive advances in the future.

The challenges of HRI are inherently interdisciplinary, and drawing upon multiple contributing competencies is necessary to have successful, relevant outcomes. For example, the investigators in this area have, for the most part, backgrounds in psychology, and they could benefit from the knowledge, alternative frameworks, and approaches of experts in robot navigation, control, and modeling and simulation. Multidisciplinary approaches can interrogate issues from multiple perspectives, promote the integration of insights, and facilitate the connection of ideas in novel ways. Over the years, ARL has apparently supported a body of work on robotic control and navigation and already has many experts on staff, including mathematicians, electrical engineers, mechanical engineers, and physicists. There now appears to be a great opportunity for joint development of models and algorithms supporting this work. The description of research in multiagent systems appeared to focus on the human factors aspects of the systems and did not describe collaborative interactions with robotics researchers. Such collaborations are essential for a full understanding of the robotics features of such systems.

There is a wealth of available knowledge to be leveraged on how robots identify where they are, map their environments, plan their movement in autonomous or semiautonomous settings, use beacons and milestones, identify obstacles and hazards, fulfill commands, navigate, and follow trajectories. There is an opportunity to use this knowledge in the design of joint human–machine action. Missing out on this wealth of knowledge means that the algorithms developed by the Human Sciences Campaign may miss out on a variety of behaviors and constraints that are well studied and well modeled by individuals with mathematical, mechanical, electrical engineering, and computer science training.

Another area where ARL research on joint robot–human action can benefit is the significant body of work on hard fusion. Also of increasing significance is the emerging field of soft and hard fusion. Soft and hard fusion appears to be a major opportunity for developing a framework for HRI scene interpretation, because it admits a large suite of heterogeneous inputs, including heat sensors and tweets. There are numerous individuals across ARL with relevant expertise in this area, but there was scant evidence of any collaboration with these individuals.

ARL identified a focus on sociotechnical systems, including data to decisions, decision support systems, human dynamics of cybersecurity, and network team performance. Although ARL identified human dynamics of cybersecurity as a component of ARL's focus on sociotechnical systems within its Human Sciences Campaign, ARL's Human Sciences Campaign did not present anything that was germane to cyber research. Given that there exist many opportunities for adversaries to compromise robots or influence people, this would seem a critical area for human sciences research. This work may already be ongoing under a different ARL campaign, but it is important to support proactive collaboration and engagement of the human sciences in this area.

The research on understanding sociocultural influences suffered from some serious methodological design flaws and weaknesses in analysis that completely undercut its potential usefulness and impact. Chief among the problems was the study's reliance on a subject population that was not representative of the target population. More specifically, subjects were uniformly Christian, white males who did not remotely reflect the religious, ethnic, racial, or gender demographics of the Army personnel targeted by the study, making it difficult and misleading to draw any credible conclusions from the findings of this study. The serious problematic design of this study suggests a need for preapproval quality vetting of proposed research designs at ARL.

REAL-WORLD BEHAVIOR

Accomplishments

The collection and analysis of human behavioral data within dynamic, complex, natural environments is an ambitious and challenging undertaking. Not surprisingly, the accomplishments in this area are only somewhat incremental given the immature state of the art and the groundbreaking challenges to develop needed enabling technology and methodology. Continued strategic investment to push advances in this area would yield significant payoffs for the Army and potential spin-offs of benefit to other government and private sector research and development.

Both the environment for auditory research (EAR) facility and the SPEAR obstacle course are outstanding, world-class facilities that have been brought in to this key campaign initiative. The research team using these facilities is well focused and has effective leadership. Ongoing hardware updates to the EAR facility will ensure that it is state of the art and easy to use for advanced studies of auditory perception. Since the ARLTAB's 2013-2014 review, the research at the EAR facility has become more general and more relevant to the real world. Specifically, studies appear better designed, are less controlled than traditional psychoacoustic work, and address questions that can impact situations beyond the specific conditions tested. These advances could beneficially be pushed even further to raise the ante on research outcomes to an even higher level of importance and excellence. The collaboration with the neuroscience group to jointly measure behavior and neural signals, using electroencephalogram (EEG) technology, is laudable.

The research presented in this area appears focused on mission-relevant problems and contexts and draws on measures from multiple domains (e.g., biomechanics, cognition, and neurosciences), consistent with the goal of addressing real-world complexity. For example,

- The research using electrocortical activity to distinguish between uphill and level walking is apparently the first study to demonstrate that cortical activity changes while the human walks over different terrains. However, given the limited statistical power of the EEG signal to distinguish the terrain condition (level or incline), it would be necessary to consider more sophisticated analytics (e.g., source analyses for EEG studies) or other methods (e.g., signal enhancement methods that can reduce artifacts from electrode movement) to provide feed forward control signals for exoskeletons. Regardless of outcome, this type of work is a good example of the foundational research needed to support the development of highly mobile sensing systems that could be useful in the field. Advanced measures for evaluating signal quality need to be developed, and the translational neuroscience (TN) group needs to ensure that it is aware of and understands the lessons learned from prior work in this area.

- The work on stretchable conductive elastomers for soldier biosensing was impressive and, if successful, could have applications well outside the military realm. This is also important foundational work, showing encouraging progress, to enable EEG measurement under real-world conditions where soldiers are moving, sweating, and otherwise burdened.
- The effort showing the effects of marching, rucksack load, and heart rate on shooting performance in the field is an applied study with considerable generality, because it is designed to understand the effect of work and fatigue on human performance. The study makes excellent use of the SPEAR facility and is very relevant to real-world Army combat, where soldiers may carry heavy rucksacks for extended periods of time.
- The research dealing with temporal and semantic coherence of sounds has implications for the presentation of multisensory data for training—for example, the possible effect of latency in presentation of initiating event and audio and visual stimuli generated in response to the event.
- The research on a novel measure of driver and vehicle interaction and research on eye movement correlates of behavioral performance in a simulated guard duty task showed interesting use of validating measures other than EEG. The novel driver–vehicle interaction metric and eye-tracking measurement used appeared less complex with respect to the collection and interpretation of data than the EEG measures that seem to dominate ARL's real-world research paradigm.

Challenges and Opportunities

Measuring and understanding behavior in the real world is an audacious objective given the limits to the current understanding of nonlinear causal propagation and emergence in complex real-world systems and environments. True multidisciplinary projects are generally rare but are essential in order for this area to mature to yield meaningful and valuable outcomes.

Research in the real world requires giving up experimental control for the most part. It requires ecologically valid contexts and tasks and the means to capture and analyze the nonlinear causal interactions of multiple variables that may contribute to variance in performance and behavior. For example, it is not appropriate for real-world studies of auditory localization to assume fixed head and sounds unnaturally short that are turning on or off simultaneously. Expertise in ecological psychology and complex sociotechnical systems engineering is needed and could provide useful insights and research strategies under this initiative.

It may be possible to conduct some tightly controlled studies as a tool for assessing whether a new technique is feasible and then progress to a study in a more ecologically valid setting. Indeed, there may be a continuum of increasing complexity within which it may be possible to effectively derive valid causal assumptions. Therefore, based on findings, the driving–vehicle interaction work might be repeated for validation in more capable simulators before moving out to data collection in real-world driving in the field. Developing a methodology for moving research between the laboratory and real-world settings would be a major step forward and a significant contribution to establishing a methodological advance for real-world research.

The project on systems-based metrics of team states using communications data is somewhat ambitious, envisioning the collection of massive amounts of data on hundreds of individuals on a 24/7 basis on brigade-level mission command training exercises. Big data analytics will then be applied in an effort to explore the extent to which group efficacy is a function of variables such as shared understanding of command intent. The hope is that the data collection will be unobtrusive and that it will permit intervention before an existing or potential problem. The analysis of this massive volume of heterogeneous team communication data appears daunting and might benefit by collaboration with information technology or computer sciences subject-matter expertise resident elsewhere in ARL.

TOWARD HUMAN VARIABILITY

Accomplishments

Understanding and predicting human variability is an important and timely topic of investigation. Current systems are calibrated to the average performance of the average person in challenging circumstances; optimized adaptive systems might enable better use of human capacity when situations and states permit. Advances in this area reflect the availability of increasingly sophisticated behavioral and brain measures and the development of new analytic and statistical methods that may enable adaptive systems in operational settings.

The human variability research at ARL addresses the variability in performance between different individuals and the variability within one individual at different times and particularly in different states. Two components of this research initiative were presented: (1) cognitive neuroscience research and development with a focus on real-world measurement designed to illuminate the connection between brain states and behavior and their potential implications for brain–computer interfaces and (2) a proposed initiative to understand and predict variations in performance with measures of human behavior and physiological state over long periods, using relatively unobtrusive immersive measures interspersed with laboratory-based tests and measures.

Overall, the work in this area seems to be of exceptional quality. The majority of the research efforts used EEG (or in some cases fMRI) indexes of brain responses associated with visual or auditory processing or with motor responses. Several of the newer projects used either EEG sensor data or fMRI activity to identify patterns of coactive brain circuits that are characteristic during different stages of sensory, motor, or cognitive tasks. Novel work was presented aimed at advancing the state of the art of dynamic analysis of brain signals to identify active brain circuits and show how activity within these brain circuits differs within individuals and between individuals across the time course of responses in typical operator-relevant tasks. The research infrastructure in EEG, eye movements, and (through university collaboration) fMRI is superb and well suited to address the target issues.

This group has recruited an exceptionally strong set of researchers, including well-qualified postdoctoral and early-career scientists representing different technical backgrounds, with some gender and ethnic diversity. A number of the presentations featured early-career scientists. Overall, the presentations indicated good mentorship of these early-career researchers.

The quality, productivity, and scope of coverage in journal publications and presentations continue to be exceptional. The recent peer-reviewed publication record indicated between 5 and 10 journal articles per year since 2012, including 6 to date in 2015, with 10 additional articles submitted. The focuses of publications are well distributed over the scope of the initiative and are directed at cutting-edge issues in human variability. For example, a number are focused on EEG and/or survey methods (e.g., the Big Five inventory) and a number evaluate intra-individual gender measures. The majority of the verbal and poster presentations for current and completed work were focused on brain states and behavior in a variety of operational situations.

An informal network of shared resources and shared knowledge is developing that transcends organizational boundaries. In particular, there appear to be considerable cohesiveness and teamwork that extend across ARL laboratories. The level of awareness and cooperation across the laboratories is impressive, the resources needed to advance the work are available, and the collective expertise can rival that in top academic institutions, where a more insular attitude is commonplace. Cohesiveness can be a powerful tool, and ARL deserves kudos for maintaining this excellent environment.

Challenges and Opportunities

Many of the projects presented appeared at first to be at formative stages at both empirical and theoretical levels, and the range of problems being worked on seemed somewhat vague, diverse, and unconnected. However, the implicit focus was clarified somewhat during small group discussions with the research staff, who described the focus as the identification of potential brain–behavior signatures for brain–computer interface translation.

The development of practical adaptive systems based on assessment of operator state is an important research objective and would benefit from a focused set of research priorities using a well-defined set of high-value operator tasks and operational contexts. The current effort appears devoted to testing and gaining expertise with various biometric measurements (e.g., field testing of EEG and multimodal sensors). However, the diverse array of biosensors and the extent of potential behavioral components require some winnowing of possibilities. In addition to high-fidelity brain measurements examined in the laboratory, a range of other measures such as multiple cue or cue fusion approaches needs to be considered, with the goal of identifying potentially lower fidelity neurocognitive measures of brain state. Fused cue approaches use multiple data modalities (e.g., heart rate measures, eye movement behaviors, simple EEG indicators, and other behavioral indexes) to achieve possible benefits of cue integration in the classification of human states (e.g., attention, vigilance state, fatigue). This could, in turn, be foundational to the notion of precision performance, whereby inputs, decisions, and scheduling demands can be tailored to the individual, including the mental state of the individual.

One caution is that the ARL studies on individual differences focus almost entirely on developing neurophysiological measures, with little or no incorporation of more traditional behavioral and psychological measurements. As a counterexample, ARL's reported research has shown that a relationship exists between the spatial abilities of operators and their performance on robotic tasks. This finding illustrates how much information on individual differences might be missed if the focus is entirely on biological sensors. Advice on the selection of appropriate psychological measurements depends on the parameters of a particular targeted behavior (e.g., precision), but their incorporation could be given a careful and more complete consideration in the overall research strategy.

The planned new initiative in unobtrusive immersive measurement of behavior is potentially an enormous problem domain. The findings and best practices from other related major data projects need to be leveraged to identify the research niche with the highest potential return on investment for ARL. As an example, there is a program of the U.S. Department of Transportation designed in part to characterize state precursors of traffic events. Other potential opportunities for leverage include advances in detecting and correlating attention or fatigue states with human performance in smart-home projects or monitoring system efforts associated with medical issues or adaptive support for aging populations.

One of the core challenges faced by this work—as in other initiatives addressed earlier—is the availability of suitably representative human subject populations for research testing. These might include civilian populations where those are appropriate or selected access to military personnel, where data need to be representative of the populations to which the research is expected to be applied. Suitably representative populations are especially important for credible research on interindividual variation.

Another core challenge is the need to engage appropriate information systems and computational resources where needed. The scope of data that could be collected in the immersive measurement for understanding and predicting individual variation in performance has the potential to be massive. The measurements planned in the immersive workplace (possibly including recording facial expression, direction of gaze, all keystrokes in work projects, and some simple physiological measures of heart rate) suggested the generation of large amounts of data. While preliminary and pilot testing of such data in smaller units may be feasible without special arrangements, the large data demands of ongoing immersive projects of this kind are likely to require specialized data management plans and adapting existing algorithms or developing new algorithms for data mining and data analysis. This would necessitate engagement of more expertise in information systems and computational resources. While the information systems and computational resources were not part of the research program under review, it was indicated that arrangements were under way for broader collaboration with other ARL researchers and staff with expertise and resources in this area.

The cognitive neuroscience group has recruited a number of early-career researchers and collaborators from academia. Although there was evidence for strong mentoring of these individuals, the expansion of the group, combined with the fact that key individuals are moving up into administration, suggests a need to develop expanded procedures for mentorship and for the identification of a set of mentors.

For the planned research into characterizing and predicting human variability, a series of workshops is being planned to discuss relevant research topics to enable program development. This is laudable, and a beneficial expansion of this effort (perhaps under ARL's open campus initiative) could include short-term visitors from academia or industry to broaden intellectual engagement around planned initiatives such as that for large-scale immersive measurement and individual variability.

Research on individual variability is one of the most difficult problems in the field of psychology, and there is much motivation to understand the phenomenon. High-quality and high-impact research in this area, focused on measurement, estimation, and prediction, could position the group for a broadly recognized leadership role. Unique high-impact studies of individual differences and intra-individual variability as it relates to stress, fatigue, and other psychophysiological states in targeted operational environments could also provide significant benefits, beyond those envisioned by ARL, to external university or industry research efforts that generally do not have direct access to these operational contexts.

TRAINING

Accomplishments and Advancements

Effective execution of the ARL Training program is a challenging endeavor. The technology baseline on which it depends cuts across multiple S&T areas and, in turn, requires a diverse, multidisciplinary workforce. Likewise, the relevant basic cognitive and social science research that applies to many training problems is fragmented and somewhat characterized by competing micro-theories. Additionally, the advanced training and simulation technologies are increasingly linked to commercial products and requirements with rapidly evolving capabilities.

The Training program is addressing important scientific and technical challenges that are relevant to both the Army and the larger education and training community. The presented projects are employing appropriate methods and using relevant environments and subject populations.

The program seeks to improve training efficiency and effectiveness through a combination of training technologies and methods. The projects presented addressed a number of challenges, such as how to create affordable intelligent tutoring systems, how to provide simulation-based training for dismounted soldiers, and how to increase the fidelity of medical mannequins. Such work is valuable for domain-specific applications and offers significant spin-off potential to education and training in general.

Publications are generally increasing, and there is a continuing effort to expand beyond technical reports. While additional emphasis is needed to encourage publication in top-tier journals, as opposed to conference proceedings, documentation of the group's productivity is consistent with that of similar organizations.

In many cases, affordable training technologies can only be achieved through the use of commercially available technologies or products. ARL staff are aware of such dependencies and are directly addressing them in projects involving virtual worlds, wearable displays, and lasers. Projects such as Head Mounted Displays for Augmented Reality Training and Lithographic Vertical-Cavity Surface-Emitting Lasers (VCSELs) are actively feeding technology enhancements, developed to meet Army-specific needs, back into the commercial arena.

Improving training methods and technologies requires multidisciplinary teams that include behavioral scientists, engineers, computer scientists, and training designers. In general, the research teams involved in the programs presented to the panel showed an appropriate mix of the needed disciplines.

There appears to be a proactive effort to recruit and develop a competent workforce. Early career engineers and scientists were well represented in the presented programs. The Training program seems to involve an active effort to recruit postdoctoral researchers from the local community and is also employing interns to grow the future workforce.

Training Area Organization

ARL Training R&D is organized around two broad research thrusts, each of which is divided into two subareas. These research thrusts and subareas are shown in Table 7.1. Table 7.1 also shows the seven research projects presented during the 2016 review and their location within the ARL training taxonomy.

The four subareas and their associated research projects highlight the importance of both methods and technology for training. The Adaptive Training and Education subarea seeks to improve the effective-

TABLE 7.1 Seven 2016 Training Research Projects, as Aligned within the Army Research Laboratory's Human Systems Campaign's Training Taxonomy

ARL Training Taxonomy			
Effectiveness and Training Methods		Simulation and Training Technology	
Adaptive Training and Education Subarea	Training Effectiveness Subarea	Enabling Technologies Subarea	Technology Applications Subarea
Generalized Instructional Framework for Tutoring (GIFT)	An Empirical Assessment of Live vs Virtual Training Method for Dismounted Infantry Soldier Skills Training	Head Mounted Displays for Augmented Reality Training	Simulated Human Tissue Performance
Models of Expert Performance for Support in an Adaptive Marksmanship Trainer		Virtual Human Technology	Lithographic Vertical-Cavity Surface-Emitting Lasers (VCSELs) for Sensing

SOURCE: Supplement to Army Research Laboratory S&T Campaign Plans 2015-2035.

ness and efficiency of training by developing advanced intelligent tutoring and computer-based training. Key technologies being addressed include automated authoring, learner modeling, performance assessment, and instructional management. The Training Effectiveness subarea focuses on how components and characteristics of training systems—for example, media, devices, fidelity, instructors, and method of instruction—combine to determine a training outcome and how to measure those outcomes. The goal is to develop techniques to measure training outcomes and to create models to predict those outcomes for both individual and team training. The Enabling Technologies subarea focuses on simulation and training technologies that are foundational components of a variety of training systems. These technologies include virtual, mixed, and augmented reality; computer-generated or virtual humans; and human–computer interaction with the training environment. The Technology Applications subarea focuses on applying simulation and training technologies to meet domain-specific training needs.

Generalized Instructional Framework for Tutoring

The Generalized Instructional Framework for Tutoring (GIFT) project is developing the components necessary to assess a learner's state, recommend an instructional strategy, and modify the learning environment to support adaptive intelligent tutoring using a modular, open-source architecture. GIFT represents the opportunity to significantly advance adaptive intelligent tutoring and illustrates the need for organizations to make long-term investments in key research areas. The GIFT presentation to the panel provided a clear overview of its architecture and technical status. It did not, however, provide insights into how user experience and cost savings are being captured and fed back into the development effort. ARL reported that discussions on a transition and sustainment strategy are ongoing with the Army's Training and Doctrine Command (TRADOC) and the Advanced Distributed Learning (ADL) collaboration.

The GIFT project has demonstrated significant technical achievements, and ARL staff have documented those accomplishments for the training community. It has also established a broad and diverse set of international users. The expansion into psychomotor skills training represents a significant advancement of intelligent tutoring research. The annual expert meeting is an exemplar of excellent program management. In addition, the GIFT project has taken the initial steps toward technology transition by initiating discussions with TRADOC and the ADL collaboration.

Developing Models of Expert Performance for Support in an Adaptive Marksmanship Trainer

The adaptive marksmanship trainer project is attempting to develop models of expert performance and use those models to tutor individual marksmanship skills. The project has both customer and subject-matter expert support. Successful expansion of adaptive intelligent tutoring to include psychomotor skills training offers significant payoffs across a wide variety of areas, including medicine to sports.

Virtual World Research—Soldier Training Effectiveness

In an effort to reduce live training demands for dismounted soldiers, ARL is investigating the use of virtual world technology that combines aspects of both virtual and constructive simulations. Virtual worlds are readily available game-like environments (e.g., Minecraft, Second Life, World of Warcraft, and VBS3) that run on commodity computer platforms. A unique avatar that directly interacts with the simulated environment represents each trainee. Each avatar carries out appropriate actions based on the inputs provided by a trainee using various human–computer interaction technologies (e.g., mouse, joy stick, voice). If the training research community can correctly identify the cognitive and team training requirements, it may be possible to effectively train those cognitive and team skills using a virtual world and integrate that training with psychomotor skills during live training. The project began in April 2016 and has made substantial progress. It is addressing a training and technology area that has little training effectiveness data to guide potential implementations. However, it requires cognitive science and instructional design support to help identify the cognitive components of dismounted soldier training and design appropriate instruction. Because virtual worlds and serious games continue to attract a great deal of research interest, this project needs to actively explore collaborations and capitalize on related work in industry and government, such as VBS-Pointman, as well as the avatar research and development efforts at the Institute for Creative Technology (ICT).

Head-Mounted Displays for Augmented Reality Training

Augmented reality integrates digital information into the user's environment in real time. It offers the potential to enhance live training by blending virtual objects into the live training environment. A key component technology is a wearable display to present these virtual objects to the soldier. However, training applications involving augmented reality are hindered by displays that have narrow field of view and poor resolution and that are overwhelmed by ambient light. This project is specifically addressing those limitations for live training. It has a well-defined technical approach and path and is a Small Business Innovative Research success story.

Simulated Human Tissue Performance

This project seeks to develop simulated tissues that more closely approximate the mechanical and material properties of actual human tissue. Current work is quantifying the characteristics of human tissue and identifying the differences between those human tissues and the simulated tissue used in a variety of medical mannequins. The assumption is that use of more realistic tissues will improve the effectiveness of simulation-based medical training. While this assumption may be correct, no training-need analyses indicating that the fidelity of tissue simulation leads to negative training or limits the capability to adequately train specific procedures were presented to the panel. The project has established strong collaborations with several medical schools, and it might be worth exploring opportunities for collaborative research on tissue properties and manufacturing with ARL's Weapons and Materials Research Directorate.

Lithographic Oxide-free Vertical-Cavity Surface-Emitting Lasers

The Instrumented-Multiple Integrated Laser Engagement System (I-MILES) enables realistic engagement training in a sophisticated laser tag environment. This project is attempting to develop new lasers that will reduce cost, increase both accuracy and reliability, and improve eye safety. It has a clearly defined technology readiness level as a goal and a path forward that includes planned collaborations with the National Institute of Standards and Technology, the Defense Advanced Research Projects Agency (DARPA), and industry.

Opportunities and Challenges

This was the panel's second visit to review ARL's training and simulation R&D in Orlando, Florida. In 2012, the Simulation and Training Technology Center (STTC) had merged into the ARL Human

Research and Engineering Directorate. At the first onsite review of this area in June 2013, there was broad consensus by the panel that the human sciences were not having sufficient influence in areas of R&D where they could be drivers. With the notable exception of the intelligent tutoring area (GIFT), the extent to which this may still be the case was somewhat difficult to determine during the current review, because the programs sampled were generally focused on training technology rather than training science. For example, the adaptive marksmanship project was concerned with automated performance measurement. No real attempt was made to tie the work to a behaviorally based view of psychomotor skills training. The virtual worlds project is currently focused on technology and interfaces. Behavioral support to help identify cognitive components of dismounted soldier skills is supposed to be provided to the program in fall 2016; the synthetic tissues project is a technology project without any upfront training analysis. The assumption is that higher fidelity is better and that current data shows that synthetic tissues do not match the properties of human tissues. Augmented reality is a technology program, but it is addressing human performance limitations involving wearable, see-through displays. The behavioral components (e.g., psychophysical and instructional design) remain unclear. The laser project is a technology project focused in improving cost and performance for I-MILES. The absence of an overall roadmap and an explicitly defined sampling strategy, combined with presentations that were not particularly well crafted for the purposes of the review (lacking clear description of project goals, methods, data analyses, theoretical underpinnings, and connection with relevant extramural research), made it impossible to fully understand the behavioral, cognitive, or instructional components of the overall training program.

The GIFT project has the opportunity to become a de facto or de jure standard for adaptive, intelligent tutors. The continuing growth in users and the integration of psychomotor skills training represent expanding opportunities. However, to achieve its full potential, the GIFT project could continue to evolve broadly applicable models based on empirical research and formally document strengths and weaknesses as identified by its community. The GIFT project needs to define what constitutes scientific and technological success and establish a clear transition and sustainability path that includes maintenance of standardized interfaces and models independent of the core ARL S&T program. ARL also needs to develop a plan to maintain a research version of GIFT that will enable the expansion of intelligent tutoring into new domains and needs to support other training research efforts.

Integrating virtual simulation into live training for dismounted soldiers involves significant technical challenges. Virtual simulation (i.e., real people operating simulated systems) is a well-established method of training individuals and teams across a variety of skills, including psychomotor, decisional, and communication skills. In virtual simulation the individual's interaction with the simulated environment is typically accomplished through the use of simulated equipment or systems (for example, gunnery and aircraft simulators). When an individual interacts directly with the environment without mediating equipment or systems, it becomes very difficult to create an effective human-in-the-loop simulation, because virtual simulation technologies do not allow users to directly grasp virtual objects or move unconstrained within large virtual spaces. Development of both augmented reality and virtual world technologies, combined with the development and evaluation of training strategies, offers potentially effective and affordable simulation media for dismounted soldiers.

The virtual human technology project conducted at the ICT at the University of Southern California has developed several avatars with realistic nonverbal cues to support interactive training across a variety of areas (e.g., junior leadership, interpersonal skills, and sexual harassment and assault prevention) and has developed a Virtual Human Toolkit. The Orlando training research portfolio has projects that include the use of human avatars, but no specific links were identified between the ICT technology and ongoing or proposed research projects in Orlando.

Department of Defense laboratories such as ARL have a unique opportunity to foster the expansion of fundamental knowledge in the sciences of learning and training. The application of fundamental knowledge to real-world training problems provides a means of identifying the strengths and weaknesses in our understanding of how adults learn, retain, and transfer knowledge and skills to address complex problems. In order to advance the sciences of learning and training, it is critical that both the successes and failures in applying that knowledge be communicated across the tech-base community. This requires continuing interchanges between the service laboratories themselves and the academic research community through attendance at scientific conferences and increased publication in referred journals in disciplines such as computer science, education, human factors, and psychology. It appears ARL could benefit from improved collaboration with universities for enhancement of their training programs. The implementation of on-line educational programs throughout the country has changed the learning paradigm for many different groups of students and employees.

The scope and complexity of training S&T is massive. As a result, it is difficult to present a unified picture of the area and of ARL's ongoing S&T programs. Although the read-ahead material for the panel provided a great deal of useful information, it did not provide a unified roadmap of the portfolio, show the relationship between various research projects, or identify criteria for success and potential exit ramps. In addition, individual projects did not consistently identify key scientific and technical milestones or estimated project duration.

INTEGRATION TECHNOLOGIES

Accomplishments and Advancements

The integration technologies technical program within the Human Sciences Campaign focuses on supporting improved and effective bidirectional integration and mutual adaptation of soldier-technology systems for robust and optimal functioning in diverse operational environments and conditions. The goal of this campaign is to integrate operational systems and technological devices to optimize outcomes. The objective for this initiative is to close the loop between multiple systems and between systems and humans to enable the creation of hybrid human–system teams suited to demanding and unpredictable environments. The topics for the present assessment focused on R&D on closed-loop behavior that included efforts in cybernetics and brain–computer integration.

Optimized performance of hybrid teams of humans, systems, and technology requires a characterization of human capabilities, system functions, and human–system interactions, together with a deep understanding of the environments within which they could operate. The Human Sciences Campaign at ARL is actively pursuing research to more fully understand human capabilities and ways that systems can capitalize on or augment them. The long-term goal includes the development of systems with environmental awareness and the ability to analyze and understand the state of the human operator to allow the development of adaptive systems for individuals. An intermediate goal is to understand the range of variability and state-dependence of human performance and to use this to design systems that could adapt to the expected variations in performance.

This initiative includes research activities focused on measuring brain states in different tasks and environments, coupled with computational approaches for analyzing this information and integrating it with other machine-based computations. This may include specifying how multimodal interfaces may assist in achieving these aims. The integration program research team identified two topic areas for integration technologies and the emphasis on closed-loop behavior—cybernetics and brain–computer integration.

The topic of cybernetics brings a newly identified structure to the processing of closed-loop systems in which states and outcomes of the human system feedback to alter the behavior of technological systems. Cybernetics is a classic area of systems and control theory that focuses on communication and systems control; there is a broad and long-standing literature in systems engineering that may prove useful in understanding how hybrid human–technology teams operate. The topic of brain–computer integration relies more directly on the primary research on cognitive, affective, and physical brain states and their role in the development and coding of multimodal interfaces.

ARL's integration program has achieved several unique and high-quality technical accomplishments. The team has defined a program and approach focused on characterizing and understanding human state (cognitive, affective, and physical) from neural and behavioral measurements. This effort can enable improved human–machine integration and interaction. They have organized and participate in a high-quality technical research community, including able junior research scientists, with strong publication records. They have been using state-of-the-art electroencephalography (EEG) methods to analyze brain state; augmenting the analyses by models of neural substrates and augmented other modalities. Staff successfully participated in the 6.1 refresh effort that has led to the creation and execution of a cohesive research program.

The integration technologies program has employed quality research methods in physiological and behavioral signal measurement and analyses, where they are using advanced machine learning techniques. The use of EEG for human state estimation was noteworthy. This broad and unique effort to assess and identify variations in underlying state includes basic research on models of neural substrate and the identification of pathways to real-world applications.

One project is a compelling exemplar of these activities. This is the project on the design, development, and demonstration of a multimodal human–system image analysis system. This demonstration project combined computer-image analysis as one expert system with EEG responses of humans searching for targets in rapid sequence to achieve optimized target recognition from high-throughput image sequences. Using deep learning neural network models (hierarchical convolution networks) of human brain responses and computer-image analysis, the goal is to combine both sources to achieve maximized target detection, a task with clear military-relevant applications. This project combines basic research on single-trial EEG, network learning models, and information fusion. The natural combination of implicit measures of human perceptual and cognitive processes alongside automated machine processing of images was an example of innovative progress in this area.

The scientists and engineers contributing to the integration technologies program include a strong neuroscience group with special expertise in EEG measures of brain activity coupled with access to other relevant measures such as fMRI at affiliated universities. The research team includes new and junior scientists with expertise in related statistical and neural network modeling with significant potential to model brain behavior in these measures. This includes important strengths in statistical modeling of EEG with generalized linear models and new developments in analysis of single-trials (relevant to current brain state) using deep learning neural network techniques.

The research team is highly qualified and focused. The research stakes are substantial, and the gains made by this group show a strong positive pace of change. The team has demonstrated research productivity with a strong publication record.

Opportunities and Challenges

The integration technologies program has the potential to significantly advance basic research and provide key information to inform the integration of the human into hybrid human–technology teams.

The program overview projecting the cybernetics approach was only recently initiated and has the potential to further inform this initiative, especially in relation to the technological systems components of integration technologies.

The integration technologies team seems to have assembled an active and expert group in EEG, statistical and neural network modeling, and for modeling of neural substrates of behavior. The opportunity exists for development of sub-goals that will move the research agenda in the direction of closed-loop systems for individuals, or of intermediate systems that could assess the likely operational capabilities of humans in certain environments, and adapt the systems to the range of performance in those environments. Another opportunity might be to integrate the more classical and modern developments in cybernetics and control theory to augment models of humans and systems. Presently, much of the modeling work on identifying brain states and markers in EEG is utilizing strong statistical methods, general approaches to regularization and cross-validation, and relatively new methods in deep learning.

In general, there were some disconnects between the overall presented program goals and the projects presented. For example, the introductory slides discuss integration of new technologies for soldiers, but new technology was only represented in a single project.

A number of opportunities for further development and associated challenges were identified. EEG as a measure offers a number of advantages (such as ambulatory measurements, a domain of particular expertise within the team), but EEG-based measurement may not be the best tool for all of the projects on which it is proposed to be used. The creation of generalized EEG models that work with the majority of individuals may not be possible. This provides an opportunity to develop novel techniques and algorithms to either generate several different models that work in different brain states or to develop truly individually adaptive measures and systems. The opportunity for collection of other state measures needs to be considered; those may be field-ready and are critical to understanding EEG-based signals. These constructs include, for example, pain, depression, multiple constructs of stress, and cognitive load. EEG data collection is important for performing state inference, but a multimodal view needs to be adopted on the signals that are gathered for this purpose. It is critical to know what other signals (e.g., psychophysiology, movement, and temperature) need to be involved to augment or provide proxy measures as the work moves forward. The program would benefit from considering incorporating transcranial magnetic stimulation in the investigation of the central nervous system adaptation to random noise and vibration.

There was evidence in the material presented for some closed-loop behavior work (i.e., the tantalizing use case possibilities to address problems related to post-traumatic stress disorder and suicide); there is significant work that still needs to be accomplished to address these challenges. ARL noted that the cybernetics project within the Human Sciences Campaign is in its early stages. The project would benefit from a strong theoretical framework and specific use cases that will take the proposed project through the entire research cycle. There is significant literature in systems science, including in estimation and control (and associated fundamental notions of stability, observability, and controllability), which are topics central to the proposed studies in closed-loop behavior.

While there are interactions occurring between the researchers and soldiers (the primary users of any systems developed), the research needs to consider additional contacts to inform and improve the ecological validity of the projects. A systematic mechanism for feeding back information from customer experience into refining scientific studies would offer many benefits.

The cybernetics research program would also benefit by developing productive partnerships with system science experts, including the mathematical modeling of coupled human–machine systems, and the definition of compelling use cases to test and validate scientific methods (e.g., use cases could be drawn from the augmentation program).

AUGMENTATION

Accomplishments and Advancements

The augmentation program team is developing and enabling technological approaches to enhance soldier performance across wide-ranging Army-relevant scenarios. Augmenting the human with technology in order to improve performance has been in existence for centuries (e.g., eyeglasses). Warfare-specific augmentation, likewise, has been available for years (e.g., rifle scope). However, the explosion of technology in recent years has afforded the opportunity to provide the warfighter with viable augmentation options.

The benefits to be derived from developments in augmentation hinge on the balance between augmentation gains and augmentation burdens. The scientists and engineers working in this area demonstrated keen awareness of the trade-offs between the burden of additional gear, machinery, weight, and maintenance against the expected benefits from augmentation devices and algorithms. For example, soldiers are already burdened by enhancements in body armor, weapons, radios, and batteries that collectively weigh in excess of 75 pounds, and there are cases in excess of 100 pounds; the additional gains that any proposed enhancement could provide needs to be anticipated and quantified to ensure net benefits.

There is appropriate focus on issues such as long-term impact on soldiers who use augmentation devices and methods and the need to monitor withdrawal effects when augmentation is discontinued after prolonged use. A potentially important issue not fully considered deals with how the ethics involved with the implementation of augmentation are presented, monitored, and managed at ARL, beyond the standard internal review board process. Of the technologies presented, it would seem that there are significant ethical issues that need to be addressed with electrode-based stimulation of the brain and deep brain stimulation.

For the most part, ARL has focused on the physiological and physical ergonomic implications of augmentation, as opposed to cognitive and behavioral human factors. Wearing a suit may not only affect metabolism but also cognitive constructs such as situation awareness. The exclusion of cognitive constructs at this early stage of the work is understandable, but to be relevant and credible, ARL will need to address cognitive and other human–systems integration factors early in the development process.

Efforts to generate standardized methodologies and metrics are important. Conceptual graphs were presented that depicted the transition from a state of degraded capabilities to that of enhanced capabilities. As the research in this area progresses, there will be a need to develop and use a rigorous approach that quantifies this transition. A set of objective metrics will be needed to assess the effectiveness of augmentation approaches, devices, and algorithms. These metrics would, in all likelihood, be specific to each set of techniques. There are two compelling exemplars of this in the area of exoskeletons. Mooney et al.[1] used metabolic cost during human walking as a metric to assess an exoskeleton; Gams et al.[2] measured the effect of a knee exoskeleton on the metabolic cost of periodic squatting. The ARL augmentation team will likely need to develop and work with these kinds of frameworks.

The effort to develop metrics for the effects of augmentation needs to be complemented by development of metrics for the overhead imposed by augmentation. Often this overhead is directly and easily

[1] L.M. Mooney, E.J. Rouse, and M.H. Hugh, Autonomous exoskeleton reduces metabolic cost of human walking during load carriage, *Journal of NeuroEngineering and Rehabilitation* 11:80, 2014, https://jneuroengrehab.biomedcentral.com/articles/10.1186/1743-0003-11-80.

[2] A. Gams, T. Debevec, T. Petric, and J. Babic, Metabolic cost of squatting using robotic knee exoskeleton, pp. 184-190 in *RAAD 2012: 21st International Workshop on Robotics in Alpe-Adria-Danube Region: Proceedings*, Edizoini Scientifiche e Artistche, Torre del Greco, Italy.

quantified (e.g., the weight of the augmentation device). The trade-off between augmentation benefit and augmentation overhead needs to be expressed in terms of these metrics, and where applicable, using ratios of physical quantities. For example, if one studies a device that increases the blood oxygen saturation level of its user, one would use a ratio such as (SpO_2)/(device weight).[3]

Distribution of data sets is a valuable research contribution. Shared databases that allow researchers from around the world to test their algorithms under common conditions have been instrumental in moving the state of the art in many fields, such as speech recognition (e.g., the TIMIT database); visual object recognition (e.g., PASCAL and MNIST databases); and EEG analysis (e.g., PhysioBank, PysioToolKit and PhysioNet databases). Similarly, the database developed by ARL can, in principle, be as significant as these databases in providing a unique set of data. In addition to aiding other research efforts, this database could optimally be collected by the ARL team and provided to various branches of DOD to test the efficacy of their practices and techniques. This could contribute to establishing ARL as a technology leader in the field of augmentation.

The distribution of these data sets would be an important professional service as a reasonable short-term trade for the augmentation team's relatively low productivity in journal publications. As the augmentation team and its research mature, they will need to increase journal publication while maintaining the distribution of these data sets. Establishing a recognized global presence generally requires substantial contributions to the open literature.

The development of physical augmentation is a major noteworthy accomplishment. The ARL augmentation team is well positioned to become a leading global force in the research and development of augmentation for healthy individuals. Research has led to important advances for the disabled, but research to help able-bodied professionals do their jobs is not effectively leveraging the advances in wearable robotics. While the military benefits of technologies such as those intended for reducing fatigue and stabilizing the aim of combat weapons are unquestionable, so are the significant high-value options these technologies offer other domains (e.g., robotic-assisted microsurgery).

Opportunities and Challenges

ARL's R&D in augmentation is highly relevant to Army needs and has potential to elevate ARL to a leadership role in specific areas of this work. The augmentation team has published several very useful studies that will likely be influential in the literature on human enhancement. Of particular interest and possible impact is the work on controllability of brain networks. This work has been performed using highly simplified, noise-free, linear, discrete-time, and time-invariant network models. This work could be beneficially expanded to show that the simplified models are valid linearizations of the corresponding models that are more complete (which are nonlinear and only short-term stationary) and could push further with work on more realistic nonlinear mathematical models to explore aspects of nonlinear controllability.[4]

In general, the augmentation research efforts could be advanced by consideration of the following challenges and opportunities discussed below.

Increased interdisciplinary collaboration across military, academia, and industry would better inform the exoskeleton work. The augmentation program is richly connected to other human sciences expertise (e.g., biomechanics and cognitive science), but the extent and the level at which such interaction occurs across other relevant ARL communities was simply not evident. For example, collaboration with the

[3] SpO_2 stands for peripheral capillary oxygen saturation, an estimate of the amount of oxygen in the blood.

[4] H. Nijmeijer and A. van der Schaft, *Nonlinear Dynamical Control Systems*, Springer, New York, 1990.

system theoretic communities in robotics, control, and signal processing at ARL and other external research groups could productively enhance the work in the augmentation program

Cognitive augmentation (e.g., neurostimulation) is underdeveloped in comparison to the physical augmentation work being done by ARL and in contrast to cognitive augmentation research reported in the literature. The military (e.g., DARPA) and elements of the commercial sector (e.g., automotive) have been actively supporting augmented cognition R&D for many years, producing much that can be leveraged. Physical and cognitive augmentation will ultimately have to work together for the warfighter of the future. Embracing this notion during the early stages of development would yield more seamless and effective human-technical systems.

A plan is needed to map and facilitate the growth of research into the perceptual, cognitive, and physical components of augmentation. The augmentation research team needs to determine its niche and research focus with respect to perception, cognition, and physical performance and the level at which contributions can be made. The team needs to identify a specific research niche that goes beyond evaluation. Evaluation alone does not constitute science or research. This would be the foundation of a research plan that establishes priorities, resource allocations, and the roles of individual researchers.

Modeling, using both predictive models and simulation tools, needs to be more effectively and iteratively integrated into the scientific program. Augmentation research team members expressed skepticism about the value of extant modeling and simulation (M&S) tools and their limited relevance to real-world sources of variance. Nonetheless, there would be substantive benefits for more extensive use of M&S tools and related software. These may include the JACK software (for human ergonomic modeling and simulation); the RAMSIS software (a 3D computer-aided design tool for the ergonomic development of vehicles and cockpits); LifeMOD; and Make Human (for the prototyping of photo-realistic humanoids). In the area of cognitive modeling, tools can be helpful that are related to ACT-R (e.g., Java ACT-R) and SOAR. Other relevant tools are CHREST, CLARION, and OpenCog. The panel does not specifically advocate any of these specific tools but identifies them as examples of M&S tools that could be considered to assist the development work.

Working without M&S tools could be costly and limiting, because the alternative—extensive experimentation with humans—is logistically complex and expensive. Also, researcher intuition and field tests alone appear insufficient for the efforts needed by the augmentation research team. Among the potential advantages of adopting and using such models and M&S software are improved realization of goals during design and implementation; avoidance of rework costs by uncovering human-performance and feasibility issues earlier; ability to capture visually (and store) best practices for future programs; opportunities for sensitivity analysis and parameter tuning; and comparison of approaches and estimates of performance before experimentation with humans commences.

It is generally good practice to develop systems by first articulating a theoretical framework that predicts certain behaviors, followed by simulation of the framework to substantiate theoretical intuition, and then followed by validation of the simulation and theory with physical implementation and measurements of the system. Often, there are multiple feedback loops among these three parts of the design cycle that lead to the refinement of the theory, generation of more detailed predictions, and the achievement of more accurate validation. This design flow is particularly relevant to the augmentation program because of the added complexity of having a human in the loop of the systems and interaction with the real world. To enable a better definition of ARL's targeted contributions to the field, accurate and precise modeling of the augmentation system is needed, coupled with a more theoretical, top-down view of the empirical work and clearer leverage of and reference to the extant literature.

ARL researchers need to extend the trade-off between gains and burdens by defining the procedures for using performance indices to illuminate the tipping point for various human augmentation systems.

ARL needs to consider developing numerical metrics for the gains achieved by augmentation as well as for the overhead that augmentation methods impose. ARL also needs to measure ratios between certain gains and the main overhead variables—for example, for the heart rate reducing gear device (the ratio between change in heart rate during certain physical activities as a result of this augmentation versus extra weight required by the heart-rate reducing gear). While such metrics would be specific to a certain kind of device or method, they may be usable in time to define a space of all key variables for a family of devices. This space can then be divided into regions that are labeled from highly efficient (where the ratios between improvement impacts and overhead is high) and regions that are inefficient (where the same ratios are low). Devising such partitioned performance spaces would allow a judicious decision on a tipping point (where a certain technology starts justifying its benefit by overcoming overhead). This approach would allow for systematic comparison of augmentation methods by measuring the distance between performance of each method and the boundary between the efficient and inefficient subspaces.

ASSESSING MISSION CAPABILITY OF SYSTEMS

Accomplishments

The work in this area is comprised of human-centered engineering and decision support methods, models, and tools supported under ARL's Analysis and Assessment Campaign. Most of the projects that were presented in this area are responding to important and specific Army customer needs, such as the efforts dealing with human modeling, field assistance in S&T, human behaviors negatively affecting engineering solutions, and fire suppressant effectiveness. Soldier surveys conducted by ARL to identify and characterize problems with equipment and systems used in the field yield valuable feedback that can, if effectively acted on, save lives, promote mission effectiveness, and potentially provide long-term cost savings to the Army.

Progress and advances are evident in the integration of human factors and systems engineering tools. A prime example of this is in the integration of SysML activity diagrams as input to Improved Performance Research Integration Tool (IMPRINT) models. The facilities, tools, and test equipment used in this area (e.g., the renovated obstacle course) appear fully up to date and are exceptional.

Commendable efforts are under way at ARL to advance assessment science by developing new models, tools, and metrics to support the acquisition and fielding of effective human systems responsive to emerging missions and threats. For example, improving the integration of human factors and systems engineering tools in the soldier decision framework model is a significant step forward. Advances in technology, system and mission environments, and changes in soldier roles and the nature of the work they do require attendant and anticipatory advances in assessment science, methods, and tools.

Challenges and Opportunities

The Army continues to be a DOD leader for human systems integration as an integral part of DOD systems acquisition and engineering. As such, it is an important challenge and responsibility for ARL to push the science of human systems evaluation, not just to be the evaluators of record. ARL needs to lead the envisioning and development of new methods and models that are militarily relevant to future technologies, missions, and threats and that will support the acquisition and fielding of effective human systems. As an example, ARL leverage of advances in cognitive engineering can support diagnosis and assessment of complex human systems, human technology interfaces, training programs, and work redesign with methods and tools (e.g., cognitive task analysis) that can identify the mental demands (e.g., workload) and cognitive skills (e.g., situation awareness, decision making, and planning) needed to complete a task or accomplish a mission.

Another major challenge is to improve the timeliness of human systems analysis and assessment to ensure that it is not cast aside, to the detriment of soldier and mission effectiveness, in the rush for rapid fielding due to war. The process of evaluating new systems in wartime has to allow a rapid response while not ignoring the human science factors. The problem is how to best tailor human systems processes to support wartime (rapid fielding) versus peacetime system acquisition. Enhanced use of modeling of system and mission environments needs to be considered for its potential to accelerate these processes.

ARL needs to be in the feedback loop, both sending and receiving on human systems integration (HSI) issues from system conceptualization, acquisition, and design, through fielding to prevent future problems and failures with the process and to ensure long-term improvements. For example, problems identified prior to deployment that were rectified need to be tracked to determine their impact in the field. Similarly, problems in the field that were identified might warrant changes in process to assure they are captured in the future. Field data on adherence to usage recommendations or human systems effectiveness of equipment need to be systematically fed back to modifications or redesign. This would enable continual improvement of the ARL analysis and assessment processes and their impact on future Army systems. An additional benefit would be a documented audit trail that could support validation of the cost-benefits of the process for use in future baselines.

ARL could use customer-funded projects to motivate human sciences S&T research ideas and directions and to provide pilot data. Because research is fundamental to ARL, even customer-funded projects could push the research agenda. A continual focus on using contract work to provide pilot data would enhance research and identify meaningful data or problems. If the linkage between ARL human sciences research and HSI applications were more transparent, it could facilitate investment by external operational customers in the work. For example, ARL researchers indicated strong desires to connect research-based findings with field-based operational effectiveness and customer-responsive work.

Researchers making connections with military personnel in the field to gain soldier perspective has the potential to be transformative. It can inform ARL research, motivate scientists and engineers, and lead to rapid and/or dramatic changes that can have significant impact on soldiers. Programs that enable ARL investigators to collect field data have been cut, thereby reducing and/or eliminating experiences and data that are extremely valuable to the HSI effort. Assuring these opportunities for research and assessment in the field continue and expand is a key priority. In a similar vein, the Army provides a course whereby new employees are introduced to Army systems and procedures. There would be value in making this real-world sensitization standard procedure available for all ARL civilian scientists and engineers.

OVERALL QUALITY OF THE WORK

Considering the information provided by ARL, it appears that the overall technical quality of the R&D program is good and continually improving. The strategic transformation to campaigns has gone well, and the establishment of ARL-wide initiatives, such as its open campus, is showing positive effects on many aspects of the technical work.

ARL continues on a trajectory of hiring highly skilled postdoctoral researchers, many of whom are being groomed to become full-time ARL employees. Publication in peer-reviewed journals and participation at professional conferences has continued to grow, coupled with increasing participation in professional activities (e.g., journal editing). Collaborations with peer communities appears healthy and provides ARL personnel with invaluable networking opportunities and the options to leverage quality research elsewhere. ARL's investment in quality R&D in the human sciences has increased its potential for impact on the present and future Army.

While there has been good progress at growing ARL basic science, more needs to be done. Unique contributions to science in the portfolio presented are not always evident. The key exception is the initiative on individual variability, which has effectively demonstrated the potential to make important contributions to the science of brain state measurement and individual differences. Overall, the ARL neuroscience laboratory, with its emphasis on high-quality R&D and peer-reviewed publications on par with distinguished university peers, is an exemplar on which to grow and sustain basic science across the human sciences KCIs.

The new emphasis on advancing assessment science in the human factors area of assessing mission capability of systems is commendable; it needs to be brought into balance with the assessment-for-hire field assistance in this KCI, to position ARL to effectively support the Army after next. The relationships built through good customer-driven work can be leveraged to gain support for the science and technology needed to advance this capability.

ARL may already possess the in-house expertise to effectively deal with many of the complex challenges anticipated to confront the Army after next; it is important to assure that beneficial synergies that exist across the KCIs are systematically leveraged. Three areas in human sciences might benefit from broader ARL exposure, engagement, and collaborations—big data analysis, autonomous or semi-autonomous systems, and EEG-based studies in real-time (or near-real-time) systems.

Linking onsite research to field-based applications work and developing a unique set of capabilities relevant to Army needs, military operations specialties, and contexts potentially represents a unique strength and feature of the ARL human sciences. More work needs to be done here. The natural tension between the comfortable technology pull of customer-driven work and the disruptive potential of innovating through technology push needs to be better balanced to best impact the Army after next. The KCIs are promising in this regard; it may take some time to transform the current program of work and its representation to consistently align with this new core competency structure that links expertise across ARL directorates.

Taken individually, each area of the Human Sciences Campaign (humans in multiagent systems and real-world behavior and human variability) can be cast as a grand challenge, because it is complex, multidisciplinary, and involves many unknowns, requiring multilevel focuses on theory, data, modeling, and engineering to meet stated and implicit goals and objectives. For example, research on individual variability is one of the most difficult problems in the field of psychology; the planned new initiative on unobtrusive immersive measurement of behavior is potentially an enormous problem domain. Measuring and understanding behavior in the real world is an audacious objective given the limits to current understanding of nonlinear causal propagation and emergence in complex, real-world systems and environments. These grand challenges are further complicated by the interdependence and overlap of these areas of the campaign. A coherent strategy and vision, not yet evident, are critically needed for how ongoing and planned research holds together, capitalizes on useful existing theoretical and empirical baselines, and cumulatively builds to achieve goals and objectives over the near, medium, and long terms.

The following projects stand out as particularly noteworthy in the areas of training, integration technologies, and augmentation.

The training program is addressing important scientific and technical challenges that are relevant to both the Army and the larger education and training community. The GIFT project stands out for its significant technical leadership and achievements that are advancing the state of the art and of knowledge for the military training communities. It is a model of excellence in program management and has the opportunity to become a de facto or de jure standard for adaptive, intelligent tutors. Additionally, the GIFT team has taken the commendable initial steps toward technology transition by engaging in discussions with TRADOC and the ADL collaboration.

The integration technologies program team has employed quality research methods in physiological and behavioral signal measurement and analyses, where they are using advanced machine-learning techniques. One project is a compelling exemplar of some of these activities: the design, development, and demonstration of a multimodal human-system image analysis system. This demonstration project combines computer image analysis, as one expert system, with EEG responses of humans searching for targets in rapid sequence to achieve optimized target recognition from high-throughput image sequences. Using deep learning neural network models (hierarchical convolution networks) of human brain responses and computer image analysis, the goal of this project is to combine both sources to achieve maximized target detection, a task with clear military-relevant applications. This project combines basic research on single-trial EEG, network learning models, and information fusion. The natural combination of implicit measures of human perceptual and cognitive processes alongside automated machine processing of images was an example of innovative progress in this area.

The development of physical augmentation is also a noteworthy accomplishment. The ARL augmentation team is well positioned to become a leading global force in the research and development of augmentation for healthy individuals. There are potential benefits to the Army of these enhanced performance capabilities, and there are also significant opportunities for valuable spin-offs to the civilian sector.

CONCLUSIONS AND RECOMMENDATIONS

Developing a detailed organizing strategy that includes measures of performance and that conveys the breadth and depth of ARL's training S&T program is a major challenge. However, an evaluation of the scientific and technical program requires that ARL provide the following for the ARLTAB review: a top-level view of major training gaps and a listing of key S&T challenges, including a description of how those challenges are being addressed through near-, mid-, or far-term programs of research by ARL or external organizations. Without more detailed information, the ARLTAB cannot fully address the science, knowledge gaps, and risks associated with the Human Sciences Campaign.

> **Recommendation. For its training program, ARL should develop a top-level roadmap that clearly identifies research thrusts, explicitly links research projects to that roadmap, provides the reason why a particular project was selected for presentation, and includes measures against which the performance of the program and its projects can be assessed.**

There are common problems and solutions that can be transferred across the three human sciences thrusts presented for the 2016 assessment. Examples of crosscutting themes include cybersecurity for technology under development and multidisciplinary cross-training on techniques and technology. However, the mechanisms for bringing together researchers across the Human Sciences Campaign programs and projects to identify potential areas of collaboration were not apparent. Likewise, the mechanism for communicating potential crosscutting research across ARL campaigns and synergistic disciplines (e.g., human sciences and system theorists) was not made clear.

> **Recommendation. ARL leadership should consider mechanisms and processes that support collaboration on issues and research that cut across the Human Sciences Campaign programs and projects, and across multiple ARL campaigns.**

8

Analysis and Assessment

The Panel on Assessment and Analysis at the Army Research Laboratory (ARL) conducted a review on August 8-10, 2016, at the Army's White Sands Missile Range in New Mexico. The review was on three of ARL's Analysis and Assessment Campaign programs—electronic warfare (EW), cybersecurity, and complex adaptive systems analysis (CASA).

ARL's Analysis and Assessment Campaign provides tools that increase awareness of material capabilities, assesses the survivability and lethality of Army systems, and both improves and simplifies the Army's decision making. The work in the EW program provides the analysis and assessment capability to operate in an increasingly complex, heterogeneous, and contested electromagnetic environment (EME). The work in the cybersecurity program provides analyses of Army systems that are in acquisition, or that are currently operational, in order to mitigate systems vulnerabilities and prevent future susceptibilities. The work in the CASA program is aimed at assessing multiple systems interactions, operational contexts, and networks to help the Army make informed decisions regarding the survivability of networked systems.

ELECTRONIC WARFARE

Accomplishments and Advancements

The ARL EW team is highly capable. The team has demonstrated its understanding of future threats for operations in complex EMEs. The reliance on information technology (IT) and on distributed wireless tactical communication networks will extend the domain of EW beyond traditional radio frequency (RF) radar and electro-optical sensor interactions to defeat opposition weapon and vehicle platforms. With new threats, the level of analysis and assessment needed will require more nuanced applications of EW, and related activities of electronic surveillance, electronic attack, and electronic protection along with countermeasure and counter-countermeasure developments. The team is definitely capable of performing EW analysis and assessment, instrumentation development, and test and evaluation. The team has state-of-the-art laboratories and has developed state-of-the-art instrumentation, which have been integrated with modeling and simulation tools to assess and test U.S. Army RF and electro-optical systems.

The latest simulators to test Global Positioning System (GPS) receiver operations are available to test GPS operations in highly contentious EMEs. The simulators can be utilized in both wired bench tests and in free space testing inside their large anechoic chamber. There is the capability to test anti-jam beam-forming GPS antennas, which can null reception of jamming signals, by using their free space propagation testing in the anechoic chamber. The experience of the GPS engineers, as well as how GPS receivers have been tested under jamming conditions, will support the development of approaches to perform EW on the operations of a variety of navigation satellite systems. Plans to acquire simulation tools for future testing—such as the M-Code GPS and simulators for testing receivers designed for positioning solutions—provide a useful forward path.

The anechoic chamber is a special facility that provides a sufficiently large volume to test platforms with various RF systems mounted on them. The anechoic chamber allows the team to operate and test systems in bands of the electromagnetic spectrum that are prohibited by law for general transmission operation. This facility also enables the testing of systems without radiating signals that could be detectable by overhead satellites, so as to provide signal emission security. In addition, the chamber provides the capability to inject real EMEs from other parts of the world and/or theaters of operations.

The SAGE[1] RF modeling and simulation tool has leveraged existing, verified, and validated Terrain-Integrated Rough Earth Model (TIREM) software. SAGE development has permitted the design of an RF modeling and simulation tool that meets the specific test and development needs of the Survivability/Lethality Directorate (SLAD). This tool has been integrated with experimental data to insure its usefulness. SAGE also supports experiments and testing by enabling the determination of the best locations for placing instrumentation in the test environment through RF propagation simulations. SAGE has also been integrated as a component of larger, more complex modeling and simulation tools.

The team has developed RF hardware building blocks for a suite of state-of-the-art EW radio test equipment and instrumentation systems. Developments include a digital radio frequency memory (DRFM) module that is a subsystem of several ARL SLAD EW test instrumentation systems. DRFMs record and play back received signals in real time. These signals can be used to spoof systems. They are also a programmable next generation of EW system threat that will be integrated by military opponents. The Army will need to develop techniques to counter this threat, in addition to understanding how to utilize DRFMs in their own operations. A DRFM has been integrated into an aircraft mountable pod for testing ground defense systems against this major EW threat. The pod was designed to radiate both forward and backward to double the time the system can radiate on a target as it passes. The aircraft pod DRFM has a traveling-wave tube (TWT) power amplifier in its front end, to provide a very powerful jamming signal, and will be used to test air defense radar systems. The DRFM is the state-of-the-art technology in threat simulation systems and will be for some time to come.

The DRFM module has been integrated into deployable multi-sensor electromagnetic warfare characterization systems (DMECS). This system is essentially a very-wideband spectrometer that allows for the display of collected and received signals by using a hypertext markup language (HTML) user interface. DMECS is a passive receiver and does not utilize the transmitter side of the DRFM device. DMECS enables collecting, storing, and viewing the EME during testing events and provides data for later evaluation and analyses of the systems being tested.

---

[1] SAGE is a government owned application, augmented with a state-of-the-art discrete event network simulator, NS-3.

A DRFM module has been integrated into the optimized modular EW network (OMEN) network controllable radio signal generation system. OMEN allows command and control of a network of signal generators to create a distributed complex RF test environment. The network of OMEN devices can also be used to test EW operations against RF communication networking systems. The ability to capture RF network traffic and then play back and mimic RF network devices shows the impact of a programmable radio system. It also enables the integration of EW with cybersecurity, since it is possible to capture RF-modulated waveforms and rebroadcast the desired waveform features to generate spoofed network radio traffic. This spoofed traffic can affect IT network operations to shut down, introduce malware, or perform a cyber electronic attack. The OMEN system has a flexible architecture, with growth potential to generate future and emerging threats. This level of integrated cybersecurity and EW is currently in its infancy. The building blocks have been developed to assemble and evolve more complex EW systems.

The electro-optical facilities within the EW program include laser vulnerability analyses for eye damage, optical sensor damage, detection of optical sensors, laser jamming, and threat assessments. Of particular interest for sensor and eye damage are new short-pulsed lasers that can generate a wideband white light pulse that exceeds the bandwidth of typical narrowband optical filters. The analyses and propagation of the short-pulse laser utilized commercial off-the-shelf tools—the shortcomings of this approach were defined and understood. Since the standard approaches to design filters involve employment of coatings to filter a narrow band of wavelength light, the interaction of the wideband white electro-optical signal against conventional filter coatings will not work. A completely new method is being assessed by the team to create a wideband fast-reactive filtering solution. This new approach employs carbon particles floating in a liquid filter. The carbon particles can block all wavelengths of electro-optical light. The filter is placed at the focal point of the optical path, so that the optical path through the filter has a narrow cross section. Methods to increase the carbon particle density at the focal point to adjust to wideband filter attenuation for eye and focal plane array protection are being investigated.

The team is also carrying out analyses of hostile fire detection sensing in order to detect and identify various types of muzzle flash from weapons and munitions based on their hyperspectral signature profile. These hostile fire detection systems can provide U.S. forces with the location of opposition weapons fire. When the location of hostile forces is known, implementation of directed counter fire is improved.

The team is also developing hardware simulations on methods to defeat heat-seeking infrared guided munitions. This development will examine methods to create appropriate countermeasures against these threats. This research will improve various implementations of countermeasure development, especially as the threats increase in capability to include hypervelocity missiles.

Opportunities and Challenges

There is a desire to insert analysis and assessment methods into earlier stages of technology development. When, and where, to include these inputs into the development process has yet to be determined. The implementation of analysis and assessment into the technology development process needs to be studied to determine when to carry out analysis and assessment. There is probably a sweet spot in the technology development pipeline that sets the team apart from other test and evaluation organizations. This sweet spot is probably somewhat earlier in the pipeline than where analysis and assessment is currently coming into the development process. It is worth considering that there are orders of magnitude more technologies that are investigated than will be selected for development. As a result, analysis and assessment are not worthwhile on all possibilities, and so a method to prioritize high-risk technologies needs to be developed.

Engineers need to utilize appropriate sources of threat data to support EW systems. Since the threat is changing, and technology is advancing at an incredible pace, systems and threats from open source data need to be considered—in addition to classified sources—in order to develop a broad view of implementations, tactics, techniques and procedures, and threat hardware. Commercial sources of threats are becoming more readily available due to software definable radio system kits and the new commercial spectrum management approaches that utilize spectrum sharing. Defining methods to include industrial and academic collaborations, which work to counter these evolving threats, will improve understanding of current and future threats.

The focus of the team appears to be on traditional EW applications involving radar systems against airborne threats. The integration of the convergence of EW with cybersecurity appears to be limited. The EW team needs to pursue activities that improve convergence and integration of EW and cybersecurity. Tools that will enable the convergence of EW and cybersecurity need to be developed. A tool that could enable this convergence is one that can move through the physical layer to the data link layer or medium access control (MAC), network, etc., to reconstruct communications packet messages from collected RF signals. Adding the cyber reconstruction of communications packets to a DRFM module would be game changing.

The environment of a system has been limited to the signals in the operational band of the system under test. The implementation of testing in complex EMEs needs more thought. It is possible that a strong out-of-band signal can pass through an RF front end to the first active component, causing issues with the operation of a system. This can happen if the strong signal can pass through the front end and cause intermodulation products that interfere with an intermediate frequency or baseband signal. Such vulnerabilities of real systems need to be tested.

It is important to note that DRFMs can be employed against communications links and network systems. DRFMs can disrupt layer 2 protocols by replicating network packets to perform denial of service for carrier sense multiple access with collision avoidance (CSMA/CA) MACs and upset timing in time division multiple access (TDMA) MAC networks. DRFMs in new form factors that will be available in the future include the following: active electronically steered arrays (AESA); multiple input, multiple output (MIMO) arrays; and multifunctional RF systems. These multiple element DRFM designs will be narrower in bandwidth to support array operations, and their numbers will be high since they will be distributed all over the operational area—with each design supporting specific functions and spectrum allocations. DRFMs will be the future of EW, and there will be many suppliers of all the types of systems designed for specific applications and threats. Since DRFM, or DRFM functional systems, will be everywhere, the only defense against DRFMs will be networks of DRFMs to detect and gather the situational awareness of the spectrum operations. DRFM, or DRFM-like technologies, are only going to grow in the future. Using available systems, analyses can be carried out to investigate vulnerabilities of these DRFMs against themselves in order to develop DRFM countermeasures. In addition to using DRFMs in tests against existing acquisition systems, there is also a need to consider how to defend (create countermeasures) against these future threat DRFM systems.

Additionally, the GPS research path includes future simulators. These simulators need to account for a wide range of current and soon-to-be-operational satellites. Regarding navigation, the testing of inertial navigation systems integrated with GPS receivers will require the expertise of an of inertial navigation systems engineer. This expertise is not currently present in the EW team.

Finally, using the results and data from testing to improve or develop models did not appear to be a universal process within the Analysis and Assessment Campaign. The development of models from test results needs to be considered. Providing these models to other Department of Defense (DOD) analyses and test and evaluation agencies also needs to be considered.

CYBERSECURITY

Accomplishments and Advancements

The cybersecurity team at ARL functions as a blue team for Army acquisition programs and operational units. A blue team is an organization that applies its expertise in cybersecurity to find areas of vulnerability in systems and then provides the systems' developers or operators with guidance to enable them to improve their security for the future. To this end, the cybersecurity team conducts exploratory development aimed at improving ARL's capabilities, which are applied in support of the blue team mission. The term *blue team* is used in contrast with a *red team*, which focuses only on finding and reporting vulnerabilities.

ARL has established a relatively large cybersecurity team of approximately 50 people. From the team's description of its activities and accomplishments, it appears to be competent and well qualified. The team provides valuable services to Army programs and organizations, and its services are in demand from various Army organizations. There is, however, more demand for cybersecurity services than there is the capacity to provide those services. As a consequence, there is not enough time to carry out necessary tool development and maintenance. The team is performing applied research, some of which has the potential to be beneficial to the ARL mission of assessing the security of Army systems.

The team members have completed a large number of industrial and governmental certifications, consistent with DOD requirements. The certification courses are valuable, not for the certification, per se, but for the knowledge that they impart. The certification-related computing environment courses provide good, deep dives into subjects. The security courses also serve as boot camps and provide basic foundational knowledge (especially for new hires) that is beneficial to the team members' effectiveness in conducting vulnerability analyses and risk assessments.

The team has discovered new and previously unknown vulnerabilities (called zero day vulnerabilities) in Army systems. These, and other discoveries, have enabled Army organizations to remediate vulnerabilities in developing and deployed systems. Contributions have also been made to the secure design of new Army systems by participating in cyber table-top exercises.

Analysis and assessments are carried out using tools such as SAGE and NS-3, as well as other tools to predict and assess the performance of mobile ad hoc networks. SAGE has been used to study the effects of protocol manipulation and exploitation. In particular, assessments on networks that employ the routing information protocol (RIP) have been performed. The team demonstrated good use of visualization tools to understand networks and their connectivity—along with the application of best practices—to document network assessment events in real time and play back events that occurred during the assessments. The recorded information is also used to train new staff. Approaches and exploits that have been successful in previous assessments are cataloged so they can be used, or built on, in the future.

Opportunities and Challenges

While the cybersecurity team is discovering vulnerabilities in Army systems, it was not clear how effective the team is at discovering vulnerabilities compared to other teams in similar roles. Metrics are not well established in the industry in this area; there is a need for such metrics.

While the team holds numerous industry certifications, more advanced training and development of cybersecurity professionals comes through attending industry events and conferences. The team was not allowed this year to attend Black Hat and Defcon, which are probably the industry's premier practical (as opposed to academic) cybersecurity conferences. These conferences are where state-of-the-art attacks and vulnerabilities, and the techniques used to find them, are discussed.

The cyber table-top exercises present a limited view of the security of systems under development. System security problems can be introduced at various stages throughout the system life cycle.

It does not appear that the team is taking full advantage of the vulnerabilities they discover to improve the security of Army systems. In particular, best practices for newly discovered vulnerabilities include generalizing them and seeking to identify their occurrence broadly throughout systems whose security is of concern. The focus appears to be on remediating discovered vulnerabilities, rather than on taking the next steps of generalizing them, and seeking to eliminate their occurrence. Newly discovered vulnerabilities need to be broadly investigated and, where feasible, approaches to removing similar vulnerabilities from systems need to be created and applied broadly.

To date, there have been limited engagements that required assessing embedded systems (e.g., the automotive-type systems connected via CanBus). These sorts of engagements are likely to become more important in the future. A research effort needs to be started on the cybersecurity of embedded systems because they are a major likely source of future security problems. Earlier inputs also need to be provided on security architecture and the design of systems under development where cybersecurity is a concern.

The demand for the services of the team has had the effect of limiting its capacity to develop or maintain a suite of assessment tools and apply best development practices to work on those tools. The team's software engineering practices (source code management and bug tracking) are not at the level of common industry practice. The team needs to use a standard software development environment, including a source code repository, a version control system, an issue tracker, and of course, regular backups. The same is true of any other software developed by other groups.

The team also needs more time and resources to develop new tools and to make current and new tools more effective. So far, the team has been unable to identify the needs or approvals that would be required to release non-sensitive tools to the public as open source, to be consistent with government policy. Release of these tools would benefit the cybersecurity community at large and help to improve the credibility of the team, which directly impacts its ability to recruit staff.

There were limited indications that the team conducts security assessments of ARL systems and tools. In particular, it was not clear that the OMEN system had undergone a comprehensive cybersecurity review.

Security review practices also need to be benchmarked against peer groups from the other services and the National Security Agency (NSA) in order to calibrate the quality of practices and to identify opportunities for improving practices.

Thought could also be given to setting up a cybersecurity future directions group with senior members and perhaps outside advisors. The current system of bottom-up project initiation and direction-setting builds staff enthusiasm but can easily miss areas that experience suggests will become important. Furthermore, the team needs a senior champion and advocate, both to serve as a mentor and technical leader and to provide senior-level advocacy (e.g., for important advanced training activities) with ARL and Army leadership.

COMPLEX ADAPTIVE SYSTEMS ANALYSIS

Accomplishments and Advancements

Complex interacting systems permeate many facets of Army institutional and warfighting operations. Accordingly, incorporating CASA into the ARL's Analysis and Assessment Campaign is likely to provide very valuable insights and important contributions to the Army. This is a very recent endeavor; therefore, this initial review and evaluation provides guidance for this new and important activity.

The goal of the CASA program is to develop a family of simulations and associated analyses suites to provide test beds and to support experimentation. At present, only two models are available: S4 software (System of Systems Survivability Simulation) and SAGE. S4 permits a stochastic scenario exploration, while SAGE is fully deterministic. SAGE is a useful tool for visualizing and, to some degree, predicting communication performance of agents in realistic environments; for example, SAGE was shown to contribute to simulating EW and cyber impact on a radio network. The use of the OMEN and DMECS to validate the performance of SAGE was also studied.

Three primary CASA applications are as follows: counter-improvised explosive device (IED) systems sensor fusion; counter-unmanned aircraft system (UAS) and manned-unmanned teaming; and EW threats to communication networks during a recent network integration evaluation (NIE). These projects are directly relevant, are appropriate for initial efforts, and have produced insightful results.

Opportunities and Challenges

The modeling efforts embodied in S4 and SAGE were initiated almost two decades ago and are narrowly focused on a few operational settings. It is not evident from these initial applications whether, and to what degree, emergent behaviors and/or adaptive learning are either being captured or described by using these two particular models. Furthermore, it has not been shown that S4 is filling a gap not already met by other tools, such as one semi-automated force (OneSAF); a strategic pause in development of S4 is needed until this can be demonstrated.

As the ARL campaigns begin to develop processes to enhance cross-laboratory collaboration, the Analysis and Assessment Campaign is encouraged to develop stronger ties to other parts of the laboratory. The Analysis and Assessment Campaign in its current state appears to be a SLAD campaign rather than a laboratory campaign. Additionally, the CASA program goals and approaches need to be clarified. Such systems analyses need to have a role across the full-spectrum of life-cycle challenges and include the post Milestone C-stage,[2] where the vast majority of Army costs are incurred. CASA capabilities and expertise could also contribute to other Analysis and Assessment Campaign critical campaign enablers, especially the personnel survivability and human systems critical campaign enablers.

A clear perspective of how CASA contributes and crosses boundaries within the ARL and DOD needs to also be articulated as soon as possible, with buy-in from the relevant constituencies. The current CASA capability is neither adequate, nor well positioned, to engage the wide spectrum of ARL needs.

CASA capabilities could enhance and support the ARL Sciences for Maneuver Campaign areas of logistics and sustainability. Innovative logistics systems and improved technologies for sustainment operations can have a dramatic impact on reducing the significant logistics burden that encumbers operating forces across major classes of supply (e.g., food, fuel, water, ammunition, repair parts, etc.). Encompassing the entire acquisition life cycle, especially long-persisting logistics and sustainment challenges, could provide an opportunity for engineering analysis to directly support the Sciences for Maneuver Campaign and the full range of force operating capabilities. The campaign also needs to be directed toward the Army standard force on force models.

Accurately modeling an adaptive enemy in an action-reaction cycle to better anticipate likely threat responses and rapidly counter them, or even preempt these responses, is an incredibly valuable capability. Such an approach could be incorporated into the CASA activity and complement the vulnerability

[2] Milestone C is a milestone decision authority-led review at the end of the engineering and manufacturing development phase. Its purpose is to make a recommendation or seek approval to enter the production and deployment phase. AcqNotes. Retrieved September 9, 2016, from http://www.acqnotes.com/acqnote/acquisitions/milestone-c.

assessment teams in ARL. ARL needs to devote CASA and other relevant analytical methods to support the vulnerability assessment teams. Additionally, it would offer an opportunity to incorporate a systems-of-teams perspective (the human dimension) as a particularly relevant paradigm.

Understanding and accurately replicating human decisions in conflict environments and scenarios (e.g., war gaming) has been a significant challenge for the modeling and simulation communities. For agent-based simulations specifically, obtaining credible decision logic is essential for model calibration, and ultimately accreditation for use. Acquiring and transferring this military knowledge and experience can be difficult to achieve in nonintrusive ways and provides a challenge for complex adaptive systems modeling. Possible sources for obtaining the knowledge and validating decision logic imbedded in agent-based models are the pre-command course war gaming exercises at Fort Leavenworth, the Battle Command Training Program, and the Command Training Centers (e.g., National Training Center at Fort Irwin). This empirically derived decision logic, which reflects current military doctrine and the military decision-making process, could then be incorporated into agent-based models for the purposes of better understanding and replicating tactical operations, including manned-unmanned teaming and autonomous systems. ARL needs to survey and evaluate these potential sources for extracting empirical decision logic from constructive, virtual, and live simulations used by operating forces. This empirical decision logic needs to then be incorporated into agent-based models for the purposes of better understanding and replicating tactical operations.

Additionally, consideration could be given to joining existing communities of excellence. For example, with respect to the modeling and simulation domain of complex systems, ARL could locally collaborate with the Santa Fe Institute on newly emerging concepts and methods for adaptive systems. They could also collaborate with Sandia National Laboratories for engineering processes, as well as practical applications in neural networks and genetic algorithms, which are used by Sandia's Center for Systems Reliability (CSR).

Relevant academic institutions and programs to engage include the Massachusetts Institute of Technology's (MIT's) Engineering Systems Division and their new Institute for Data, Systems, and Society. The Naval Postgraduate School has strong graduate and research programs in autonomous systems engineering, operations analysis, a unique cross-disciplinary program called modeling, virtual environments and simulation (MOVES), and annually presents an informative technology refresher update (TRU) on emerging technologies.

Current manpower levels and skills mix are insufficient in both capacity and capability to adequately support a broader CASA program as it expands across ARL. ARL needs to address the lack of any data scientists and operations research analysts within the complex adaptive systems and analysis program, as the lack of this expertise is a conspicuous shortfall that is a top priority. Furthermore, future project groups need to be supported and sustained by a nucleus of operations research expertise, including data analytics, that can guide multidisciplinary groups with the skill sets relevant to the tasks undertaken. The vital role that operations research and operations research analysts can, and need to, play in this activity cannot be overemphasized.

ARL could also consider establishing, at least on a temporary basis, a small external advisory group with the appropriate mix of disciplines and experience from the information, computational, human, and analytical sciences.

OVERALL QUALITY OF THE WORK

The ARLTAB assessment of this campaign is different from that of most, if not all, other campaign assessments because the Analysis and Assessment Campaign is intended to be more of an analytically focused, crosscutting, activity rather than being research focused. As a result, the criteria are different from those of research-focused campaigns. Nevertheless, the work needs to have technical depth, and the panel needs to be presented with material that exhibits this technical depth. Specific assessment factors for any panel visits need to be developed and made known to the staff. These assessment factors need to be guided by the general categories of analytical capabilities—capacity, utilization of analytical resources, and contributions.

Additionally, ARL needs to apply analytical resources as part of the Analysis and Assessment Campaign to better understand critical relationships and how they impact a wide range of variables, from recruiting standards to battlefield tactics to force design and budgetary allocations. ARL also needs to acquire and/or develop a comprehensive set of analytical capabilities that leverage other modeling, simulation, and high-performance computing capabilities to ensure adequate support for future Analysis and Assessment Campaign endeavors.

The quality of work for EW systems was observed to be good, and the staff is knowledgeable. Good insight, however, was not obtained on the technical quality of the modeling and simulation, experimental design analysis, and analytical science. Tools are being developed, and tests are being performed that answer the questions of the Army customers.

The cybersecurity team's presentations and discussions indicated that the team is competent and contributes to cybersecurity research. Improvements are needed, however, for the overall quality of the research to attain the level of top-tier government or industry cybersecurity assessment or consulting organizations. This observation is based both on the presence of some indications of quality (discovery of new vulnerabilities and demand for services) and the absence of others (contributions to secure design, generalization of findings, and application of software engineering tools and practices).

The challenge of complex interacting adaptive systems is critical and permeates many facets of ARL, and it is commendable that this challenge is recognized and resources are being pooled to address it. In its present form, the CASA capability is not well positioned to engage the wide spectrum of ARL needs. Developing a comprehensive vision, purpose, and plan for this activity will help design a more robust software environment and architecture that leverages other developments in modeling, simulation, and high-performance computing within DOD. Such developments will also put the team in a better position to integrate forthcoming efforts in human factors and EW (such as physics-based jamming models for more complex scenarios). A better assessment of the uncertainties associated with enemy-induced jamming or countermeasures or environment-induced interference would also be conducive to better decision-making.

CONCLUSIONS AND RECOMMENDATIONS

Many of the professional staff are native to the local area and also tend to come from universities that are local to the area. In addition, the universities that support the laboratory with research are the same as the ones attended by staff members. As a result, the diversity of experience and perspective within the staff can be limited, and this is a potential weakness. This can lead to emerging developments elsewhere being missed.

Recommendation. ARL should broaden the perspectives of the Analysis and Assessment Campaign staff members. Approaches that should be considered to accomplish this should include the following:

- **Utilize the ARL open campus initiative which will involve setting up relationships with research centers across the country.**

- **Utilize the other transaction authority (OTA) of a consortium acquisition model, which can introduce industry, as well as university participation, to increase the diversity of ideas and developments and open up the technology base.**
- **Utilize virtual collaboration and a virtual community of excellence, which may attract the best and brightest without pulling them to a physical location.**
- **Support personnel with degrees from local universities to attend and obtain a degree from a leading university in their field that is outside the local area.**
- **Join and participate in existing communities of excellence or establish new ones appropriate to the Analysis and Assessment Campaign mission.**
- **Increase engagement with the broader Department of Defense (DOD)/Intelligence Community and thereby increase leveraging of Army/DOD science and technology communities.**
- **Encourage coordination with various DOD analyses and test and evaluation communities.**
- **Increase interaction with the commercial industry for sharing of methods.**

The cybersecurity team was not allowed this year to attend Black Hat and Defcon, which are probably the industry's premier practical (as opposed to academic) cybersecurity conferences. These conferences are where state-of-the-art attacks and vulnerabilities, and the techniques used to find them, are discussed.

Recommendation. ARL should take whatever steps are required to ensure that Analysis and Assessment Campaign staff members attend the premier—practical as well as academic—conferences in their area. The cybersecurity team members should attend top-tier industry events, in particular, Black Hat and Defcon, as well as comparable hacker conferences.

To date, there have been limited engagements that required assessing embedded systems (e.g., the automotive-type systems connected via CanBus). These sorts of engagements are likely to become more important in the future.

Recommendation. ARL should start an effort on cybersecurity of embedded systems—they are the wave of the future and, hence, are likely to be a major source of future security problems.

The modeling efforts embodied in S4 were initiated almost two decades ago and are narrowly focused on a few operational settings. It has also not yet been demonstrated that S4 is filling a gap not already met by other tools such as ONESAF.

Recommendation. ARL should initiate a strategic pause in the development of S4 until it can be shown that this tool will fill a gap not already met by other tools such as ONESAF.

Current manpower levels and skills mix are insufficient in both capacity and capability to adequately support a broader CASA program as it expands across ARL. The lack of expertise in operations research analysis is a conspicuous shortfall. The vital role that operations research and operations research analysts can, and should, play in this activity cannot be overemphasized.

Recommendation. ARL should address the lack of data scientists and operations research analysts within the complex adaptive systems and analysis program. Future project groups should be supported and sustained by a nucleus of operations research expertise, including data analytics, guiding multidisciplinary groups with the skill sets relevant to the tasks undertaken.

9

Crosscutting Conclusions and Recommendations and Exceptional Accomplishments

The Army Research Laboratory's (ARL's) mission—to discover, innovate, and transition science and technology to ensure dominant strategic land power—demands an institutional culture that values and rewards foresight and farsighted vision while meeting the Army's current scientific and technological needs.

As technology marches on at an unprecedented pace, the relentless pursuit of innovation by means of an integrated multidisciplinary system approach is becoming increasingly important. Through an interconnected, holistic perspective and integrated systems approach, project synergies and spillover benefits can be optimally harvested. A concerted effort to understand future needs and to craft the research portfolio relevant to the Army of the future is the ultimate challenge as well as an opportunity for ARL.

The competitive institutional stature of the ARL, vis-à-vis other research organizations in the United States and abroad, hinges on crafting and executing a robust and focused research portfolio. The success of ARL researchers, in turn, is directly linked with their continued professional development in the workplace. Nurturing the research staff, senior and early-career, requires a continuing effort.

In this uncertain funding environment and fast-moving global technological landscape, productivity is another essential element of institutional success. The ability to shorten the research cycle from science to technology to useful product is essential to the institution's competitiveness and sustainability.

Additional opportunities will be presented by having the ability to effectively utilize technologies, commercial or otherwise, that are deemed critical to the well-being of the soldiers, eschewing the not-invented-here syndrome. A systematic, structured effort to leverage innovations from outside sources, by either complementary in-house projects or by collaborating with well-selected research partners, will enhance overall productivity.

During the 2015-2016 reviews, collaboration efforts were well demonstrated across ARL science and technology (S&T) campaigns and externally. Most of the projects were engaged in upward collaborative efforts to various degrees; this is commendable. The success of ARL's leadership in recruiting energetic, early-career talent was evident. However, a heightened excellence can be achieved by working toward a higher level of portfolio focus, project synergy, and overall productivity.

The challenge of a matrix management structure, with campaigns formed of teams addressing specific goals superimposed on a more traditional hierarchy of directorates, is a departure from the rigid structure of the past. It is designed to tackle problems by bringing together researchers with the needed competencies. This new challenging model of planning by campaign and executing by directorates, which is necessary for added flexibility, is still in its infancy. This structure provides an opportunity to differentiate between ostensibly orthogonal research thrusts.

In the act of constant innovation, ARL can continue building toward a best-in-class, forward-looking culture. To this end, researchers need to be relentlessly asking the questions, What is the impact? How can we make an impact? What comes next?

ARL's open campus initiative is poised to facilitate cultivation of a constantly innovative environment. The initiative also serves as a conduit for garnering the benefits of open innovation in search of the delicate balance between importing and exporting knowledge to sustain a competitive edge.

This chapter highlights crosscutting conclusions and recommendations. At the end of this chapter, exceptional accomplishments are identified that correspond to each of the ARL S&T campaign areas reviewed in 2015.

RESEARCH PORTFOLIO

The campaign thrusts at ARL comprise projects that reflect a broad scope of research activities. Significant impact across such a broad range of activities can only be realized through better integration, which, in turn, is dependent on an overarching and focused research vision.

ARL's research involves two broad mission elements: first, to respond to existing and anticipated threats and capitalize on recognized opportunities to protect and enable the modern warfighter; and second, to develop the necessary knowledge base, tools, and capabilities that will allow ARL to respond rapidly to unanticipated threats and opportunities. The majority of the research presented in this review period has been directed, occasionally loosely, to the first of these mission elements. However, it is in this second area that the greatest opportunities are to be found. ARL is in a unique position to assemble the knowledge base, computational and experimental tools, and capabilities into response systems that can be deployed as a rapid reaction to unanticipated threats and opportunities. This is a broad niche area ideally suited to the ARL mission and crucial to the nation's technological health.

A research portfolio efficiently aligned with these two Army needs will necessarily include fundamental research activities as well as awareness of capabilities available at university, industry, and national laboratories. ARL needs to cultivate the expertise and the ability to integrate and capitalize on these internal and external research efforts to make the Army agile and responsive. Crucial to the success of any such program is a clear articulation of the criteria used for program evaluation and project termination. A strategic approach to phasing out projects will assure that ARL remains focused on campaign needs and that technical programs and projects remain focused on meeting mission needs. An effective strategic approach also facilitates reallocation of personnel and other resources in support of new topics and objectives.

Exit criteria might consider whether (1) technical feasibility has been proven to be unlikely or has been determined to be unlikely, (2) work has been adequately addressed by others, (3) the Army requirements have been met, (4) higher strategic priorities demand reallocation of resources, (5) there is lack of technical progress, (6) there has been a shift in scientific and technical paradigm, and (7) there has been a determination that the expected benefit to the Army no longer exists.

Recommendation 1. ARL should articulate a vision for each research effort, its impact on options for the Army of the future, and the exit criteria to be used to decide when to terminate a project.

INTEGRATION OF RESEARCH AND SYSTEMS ENGINEERING

During the 2015-2016 reviews, which are summarized in this report, the ARL campaign areas did not uniformly present adequate information to permit confident understanding of how projects are integrated within and across campaigns and of the systems engineering whereby projects are conceptualized and planned from initial planning (considering the relevant theory base and the work of others) through application of results (considering potential for transition to development).

The problems faced by the Army of the future are extremely complex and will require options for solutions that are complex and multidisciplinary in conception. Success by the campaigns depends on effective leveraging of ARL's disciplinary competencies. Deliberate inclusion of interdisciplinary interactions in the early stages of problem formulation can yield unexpected results and may result in solutions that have a greater impact.

To optimize the progress of their research, to set a path forward for each project, and to perform tests, analyses, and experiments that produce meaningful results, researchers need to consistently analyze data and contemplate the theories that are behind the observed physical phenomena, test data, and modeling systems. A key consideration in data collection and analysis is face validity—the extent to which a test, experiment, model, or analysis measures and examines what it is purported to measure and examine. A salient example of inadequate face validity is the selection of nonmilitary human subjects for study in human science experiments purported to yield results and conclusions generalizable to the military population. It is suggested that ARL survey communities in academia, industry, and other government agencies toward establishing strategic baselines for investments in interdisciplinary areas.

ARL might consider treating its evolving interdisciplinary research and development (R&D) as a challenge problem in organizational change and assigning proactive responsibility to individuals with the expertise and mandate to develop and facilitate an ARL-centric approach that leverages ARL disciplinary strengths for each campaign area. ARL might also consider adopting more systematic and formal approaches for collaborating in a multidisciplinary R&D environment.

Recommendation 2. For each campaign, ARL should address the following:

- **Examine how projects and programs are integrated within and across campaigns and how their findings feed into one another and into common goals and share this analysis during future reviews.**
- **Apply systems engineering principles and processes across the life cycle of projects.**
- **Address validation and verification across the design of experiments, modeling, tests, and analyses.**
- **Secure military-relevant subjects for tests, experiments, and field studies involving humans.**

INTERACTION WITH INDUSTRY

Industrial groups have leading positions in a variety of areas that fall within the purview of the ARL campaigns. In order to assess the state of the art within such areas, ARL ought to be in position to ascertain the capabilities within the relevant industry and the extent to which that capability is relevant for ARL's particular needs. Such a connection with industry could minimize ARL's using scarce resources to duplicate what has already been accomplished. In addition, knowledge of the state of the art in industry, in addition to academic work, provides a yardstick to assess the level of ARL's work. There are a variety of approaches that can be used. For example, ARL could actively pursue the participation of relevant industry participants in its open campus initiative and could include industrial laboratories in the sabbatical program.

Recommendation 3: ARL should undertake a systematic effort to broaden and extend its awareness of research and development activities across industry.

RESEARCH ASSESSMENT

Effective research assessment processes require (1) clear articulation of the expected outcomes for each portfolio, including key technical milestones and metrics to be used for measuring progress; (2) definition of a similar set of outcomes and metrics for each project within a research portfolio; and (3) a "closing of the loop" that documents and describes the actions resulting from the assessment process. Beyond providing researchers the ability to gauge progress and make midcourse adjustments, a well-structured assessment process provides greater visibility to collaborators and partners, especially in research that entails a systems engineering approach.

Recommendation 4. ARL should place greater emphasis and focus on a systematic assessment of its research portfolios.

STAFF DEVELOPMENT, RETENTION, AND MENTORING

The recruitment and development of competent scientific and technical staff are essential to the successful performance of the research portfolio. Recognition of success and recognition for success go hand in hand. Associated with metrics for program and project success is a formalized reward system for staff. The reward system may include such components as monetary and nonmonetary awards, internal recognition, external peer review, research freedom, and laboratory-wide recognition of stature (e.g., fellowships).

Beyond rewards, staff development includes enhancement of the individual's professional knowledge, skills, abilities, and career growth. The increasing influx of new research personnel into ARL provides an opportunity to continue the advancement of high-quality research. A significant portion of these new hires is being educated and trained in the United States in areas of interest and importance to the mission of the Army. The research culture at ARL may be different. Recognizing this cultural difference and taking action to accelerate the careers of new research hires within the ARL culture require a formal mentoring process that is effective and efficient in transforming successful researchers from one culture to a different culture (ARL).

Mentoring is more than obtaining a favorable return on investment. Effective mentoring impacts the whole person, reflecting the mentee's strengths and needs, both internal and external to ARL. The impacted mentee is then able to provide ARL with more than high-quality research—namely, an enriched culture that becomes self-sustaining and more productive. Effective mentoring enables a successful enterprise.

Retention of senior staff who are well established researchers is also important for building and ensuring a successful enterprise. Effective mentoring depends, in part, on the outstanding senior staff. Corporate memory is strongly influenced by accomplished senior staff.

Recommendation 5. ARL should develop a structured program for professional development of its research staff and assess the program for its effectiveness.

FACILITIES AND EQUIPMENT

Facilities and equipment constitute a critical pillar of the ARL infrastructure. ARL possesses an extensive suite of state-of-the-art facilities supporting all of its campaigns and essential research areas. These facilities support high-quality research, development, and support transitions at ARL; they also represent a key attractant for recruitment and for candidates to participate in ARL's open campus initiatives. While some of the facilities are new, others are aging and warrant ongoing analysis of needs for maintenance and investments in upgrading and/or replacement. Some of the facilities are seriously underutilized, and an assessment of prioritization of continued support of such facilities within the core campaign enablers needs to be conducted.

Further, ARL's vision of Army and ARL needs and goals through 2035 requires facilities and capital equipment upgrades and acquisition investments tied to essential campaign areas and core research goals for both computational and experimental needs. The vision needs to take into account the long timeline for acquisition, upgrades and refurbishment, construction, decommissioning, and dismantlement.

In the reviews of the campaigns and core research areas, ARL has not detailed adequate information to permit confident understanding of ARL's plans for maintaining, upgrading, or expanding and/or contracting current facilities or for assessing and planning the need for new facilities, staffing the facilities, and securing funding to build and thereafter support these plans.

Recommendation 6. ARL should complete formulation of 5-, 10-, 15-, and 20-year strategic plans linked to the campaign technical goals and objectives for facilities and capital equipment. These strategic plans should also include strategic and tactical plans for necessary computing resources, in particular, those needed to support classified computational needs.

EXCEPTIONAL ACCOMPLISHMENTS

The following are the exceptional accomplishments for each campaign area.

Materials Research

The biological and bioinspired materials group has an excellent track record that includes the stabilization of proteins against thermal and chemical extremes using new chemistries and methods to derive antibody-like reagents that improve upon antibody properties (specifically bimolecular recognition and binding characteristics).

The research on structural batteries using additive manufacturing combines novel fabrication methods with insight selection of compatible multifunctional elements that combine structural components with energy storage components. Experimental work is carried out concurrent with modeling studies that guide system design choices. The external collaborations are facilitated by a flexible methodology that provides easy incorporation of next-generation subcomponent materials as they are developed.

ARL has an opportunity to move aggressively to capitalize on internal and external advances in the energy and power arena. For example, the world-leading results on enhancement in quantum-well infrared photodetector efficiencies could be translated into capability demonstrators for manufacturers and customers.

In the area of engineered photonics materials, facilities and capabilities are being leveraged into compelling device- and application-driven work, especially in ultraviolet materials, infrared devices, and the device physics in each of these areas.

The high strain rate and ballistic materials program showed an excellent degree of integration between materials science fundamentals and applications, combining simulations and experiments aimed at developing structure-property correlations with advanced processing and fabrication approaches. The miniaturized Hopkinson bar and multiscale rate-dependent mechanical testing equipment, along with micro-scale sample preparation set-up for investigating polymers, metals, ceramics, fibers, and threads, are unique facilities.

The exceptional electronic materials programs are those to remotely disable weapons, vehicles, land mines, and so on from prior military actions that were left behind or provided to a former ally turned adversary. This project is innovative, well conceived, and executed to meet a unique Army need. ARL appears to be breaking new ground in this area.

Exceptional structural materials research efforts are those intended to couple modeling with experiment with the goal of producing design tools. One such program is directed toward grain boundary modeling of ceramics for lightweight protective materials. A suite of tools is being developed to permit simulation of grain boundary structure and properties under high rate loading conditions. Although these tools are being used to investigate grain boundary structure and properties of boron-based lightweight ceramics, these same tools will be applicable to study grain boundary interfacial relationships across all ceramic materials.

Sciences for Lethality and Protection

The most impressive accomplishments of the battlefield injury mechanisms program are that it has been implemented, a strong cadre of scientists is working on it, and a credible program is under way. Almost all the battlefield injury mechanisms research topics presented had a combination of computational and experimental approaches whose interplay will be fundamental to the success of this research.

In the directed energy area, the ability imparted through the radio frequency (RF)-enabled detection, location, and improvised explosive device (IED) neutralization evaluation (REDLINE) technology applied to a convoy to sweep and destroy IEDs without interfering with operations and civilian communications is a game changer. The progress in applying an old idea, harmonics detection, to solve this problem is impressive. The investigators have done an exceptional job of transitioning the hostile fire detection technology to an operational prototype. Patents to protect intellectual property rights provide the potential of monetizing the innovation—for example, for acquisition by police forces. This level of transition is probably more appropriate for 6.3- and 6.4-funded R&D.

In the armor and adaptive protection area, the R&D described showed how ARL is building on its tradition of excellence to provide the knowledge basis for current and future Army needs in protecting our warfighters. This remains a core competency that underlies Army capabilities across the entire Department of Defense (DOD).

The teams working on energetic materials synthesis and propellants demonstrated high technical competency and in-depth understanding of the technical issues and the challenges in making progress in this area. The teams are commended for making progress in technically challenging areas. The teams understand the necessity for experimental validation and are making significant progress, although more work needs to be done. The efforts at synthesis of energetics are clearly cutting-edge work and are showing results in the newly developed promising chemicals. This area of expertise is very important to the Army and to DOD in general and is commended.

In the area of flight guidance, navigation, and control, the scientific quality is deemed to be very good. This team reflects a strong understanding of field. Further, the researchers are active in key technical meetings (American Institute of Aeronautics and Astronautics) and DOD technical exchanges (the Technical Cooperation Program, or TTCP) as well as having partnerships with other research groups (Air Force Research Laboratory) and university researchers. The research approach uses appropriate use of equipment and commercial software that is widely used in industry and research projects. The modeling approach is well established and appropriate for engineering-scale problems. The technical qualifications of team members is high, both with Ph.D.'s in the subject areas and significant experience in research in this subject areas.

Information Sciences

Of the reviewed projects, some are deserving of special mention. The work on using a distributional semantic vector space with a knowledge base for reasoning in uncertain conditions represents a strong contribution. The research features a combination of statistical and machine learning methods with semantic rules for reasoning in an uncertain environment. This work draws upon the use of semantic models with a goal of augmenting a curated knowledge base by reasoning through analogies based on statistical representations. Both the ideas and the proposed methodology contain novel elements. The work is well grounded in the literature, and the researchers are aware of related efforts in the research community.

The opportunity for strong technical contributions and for differentiation from research conducted elsewhere, as well as the value proposition for the Army, lies in a mission-oriented focus to the research. In several projects, this focus and constraints, such as limits on prior information or on available bandwidth, were a clear driver for the research. The work on autonomous mobile robot exploration with an information-gain metric stood out in this regard. This project featured a functional prototype of a robot capable of autonomous exploration. This work has opportunity for near-term application, and yet it is set in an information theoretic framework that is rich enough to support the development of more sophisticated and capable algorithms applicable to potential missions.

The work on stylometry authorship attribution for source code and binaries is noted as a strong contribution, and it advances the capabilities of attributing cyber weapons to a common origin. In particular, it extends the capability of attributing source code to a common origin to binary codes as well, and it is unique among existing approaches. Similarly, the work on a resource-conserving signature system that uses the one-way properties of Bloom filters to enhance the security of stored signatures against device loss or capture was considered to be exceptional and of high practical import.

Another project on weather impacts on microgrid renewable energy ramping event modeling entails a combination of field experimentation and modeling to enable the efficient use of renewable energy by incorporating weather data into the decision-making process. This novel approach is focused on a microgrid scale and fieldwork already undertaken to test the impacts of solar utilization and solar flux measurements on power output from photovoltaic systems. The work has received external recognition, and ARL investigators have been invited to participate in multiple related scientific panels.

While not a research project in the strictest sense, the development of emulation and simulation environments like the Network Science Research Laboratory is to be commended. It has provided a powerful capability that will enable significant network and cross-coupled research investigations arising out of future ARL and Army research needs.

Computational Sciences

Research efforts in areas of quantum computing and software environment optimization are leadership-quality work that advances basic science in important areas of computing technology. The development of a threaded message-passing interface for reduced instruction set computing array multicore processors has yielded innovative solutions to the challenging problem of power-efficient parallel programming. The work on high-performance-computing scaled quantum hardware description language is representative of one of the few efforts in the area of quantum networking; it is likely to have a strong influence on the development of future systems.

Multiscale material modeling is a potential game-changing computational technology for predictive simulation in the mechanical sciences. These multiscale simulations are essential for assessing vulnerability, lethality, and effectiveness of weapons and protections systems, and the current effort demonstrates the project's utility in theory and also in practical application to software commonly used (e.g., Lawrence Livermore National Laboratory's ALE3D production software tool, used for high-explosive weapons and target simulations). This research is of high value for predictive forward analysis, and components of this work have applications in enhancing the performance of inverse analyses and quantification of the margin of uncertainty estimations.

Sciences for Maneuver

In the human–robot interaction (HRI) area, the science is technically sound, and the work is published in top journals, including *Human Factors*. The utility of the work appears to be recognized within ARL—for example, elements from the tactile feedback project will be incorporated into the next warrior experiment. The use of soldiers in experiments is commended. The move toward more realistic warfighting vignettes and more real-life simulations, which instantiate threats and hostile elements, would help establish the value of a technology in achieving a desired capability. The research presented will be shifting from one-person/one-robot studies to multiperson/multirobot scenarios. This shift in research focus is appropriate as the Army moves toward use of more complex teaming architectures. This use case also highlights the importance of providing the right information at the right time to the humans and to the robots, identified as a thrust of the HRI program.

The intelligence and control (I&C) work employs innovative approaches in developing and supporting advanced technologies, algorithms, and tools in support of the warfighter effort. This area invests in advancing the effectiveness and efficiency of its research personnel. The focus of the I&C theme is on developing software and algorithms that enable vehicles to approach a higher level of cognition, enabling the teaming of autonomous systems and soldiers. The higher-level cognition that the I&C theme focuses on is aimed at enabling autonomous assets to work in the environments of relevance to the military.

In the perception area, the work on weakly supervised segmentation for mobility is significant for several reasons: an interesting science of vision result was published at a major conference; it is an integrated end-to-end project that demonstrates the value of the research; and it involves an external collaboration with a university.

In the platform mechanics area, the Koopman decomposition of periodically excited Hopf bifurcation systems is outstanding research. The investigators have taken advanced mathematical methodology in nonlinear systems theory and applied those tools to current, meaningful Army problems. By the use of nonlinear modal decomposition, the investigators have been able to take complex experimental data and decompose them into the fundamental modes of interest, thereby giving insight into the physics of the underlying mechanisms. The method is broadly applicable and not limited to one particular type of nonlinear problem.

Also, the design and characterization of stretchable electronic materials and components as soft enablers is excellent research. This is a very ambitious project aimed at creating stretchable electronics using micron-scale particles embedded in soft elastomers to enable soft robotics. A range of particle aspect ratios—from spheres to rods—is used to address various and interesting application spaces. By tuning the aspect ratio of the microfiber, they have demonstrated the ability to achieve positive resistivity, negative resistivity, and also sharp resistivity changes with changes in strain.

In the energy and propulsion area, the experimental programs in hybrid gears are of high quality. The overall objectives of this research effort is to enable weight reduction of power transmissions, thereby increasing the power density. In order to accomplish this objective, experimental investigation of hybrid gears under adverse conditions is being conducted. This research effort is of high relevance and has been ongoing for a few years. In addition to the potential weight reduction, hybrid gears may also provide vibration and noise reduction.

In the logistics and sustainability area, the advanced sensor fusion research program is a well-thought-out analytical integration of measurable and detectable damage, and structural behavior. This provides a very tangible example of very advanced and powerful use of sensors with a clear path for in situ, in-mission use. The collaborations with airframe manufacturers and universities is particularly laudable. This research effort captured an important challenge.

Human Sciences

In the human variability research, the initiative on individual variability, with its emphasis on high-quality R&D and peer-reviewed publications is making important contributions to the science of brain-state measurement and individual differences. The recent peer-reviewed publication record indicated 5 to 10 journal articles per year since 2012, including 6 to date in 2015, with 10 additional articles submitted. The focuses of publications are well distributed over the scope of the initiative and are directed at cutting-edge issues in human variability.

In the training area, the generalized instructional framework for tutoring (GIFT) project stands out for its significant technical leadership and achievements that are advancing the state of the art and of knowledge for a broad and diverse set of international users in the training community. In particular, the expansion into psychomotor skills training represents a significant advancement of intelligent tutoring research. Also noteworthy are the proactive efforts at technology transition with TRADOC and the initiative on advanced distributed learning.

In the integration technologies area, ARL has significantly advanced the characterization and understanding of human state (cognitive, affective, and physical) from neural and behavioral measurements. These efforts provide an enabling foundation for improved human–machine integration and interaction. One compelling exemplar is the design and development of a multimodal human–system image analysis project and demonstration system. This project combines computer image analysis using deep learning neural network modeling with the electroencephalogram responses of humans searching for targets in rapid sequence to achieve optimized target recognition.

In the augmentation area, the development of physical augmentation is a major noteworthy accomplishment. The ARL augmentation group appears well positioned to assume a leadership role in the research and development of augmentation for healthy individuals. Just as existing research in this domain has led to important advances for the disabled, ARL research leveraging advances in wearable robotics will assist able-bodied professionals to do their jobs effectively in the future.

Analysis and Assessments

The Analysis and Assessment Campaign laboratories have been developed with integrated state-of-the-art instrumentation, modeling, and simulation tools used to assess and test U.S. Army RF and electro-optical systems. Specifically, the anechoic chamber provides special capabilities that ARL is well positioned to take advantage of. The digital radio frequency memory module also has a variety of potential applications and has been integrated into a network-controllable radio signal generation system that allows command and control of a network of signal generators to create distributed complex RF test environments. The electronic warfare team has demonstrated an understanding of the future of electronic warfare threats for operations in complex electromagnetic environments. The cybersecurity team has discovered new and previously unknown vulnerabilities (zero day vulnerabilities) in Army systems. These, along with other discoveries, have enabled Army organizations to remediate vulnerabilities in developing and deployed systems. ARL also has the opportunity to capitalize on such discoveries by conducting root cause analyses of discovered vulnerabilities and ensuring that future Army systems are free from similar problems. Contributions have also been made to the secure design of new Army systems by participating in cyber table-top exercises.

Appendixes

A

Army Research Laboratory Organization and Science and Technology Campaign Framework

Figure A.1 is an organization chart for the Army Research Laboratory (ARL), Figure A.2 is ARL's science and technology (S&T) campaign framework, and Table A.1 maps the ARL organizational chart to the campaign areas reviewed in 2015.

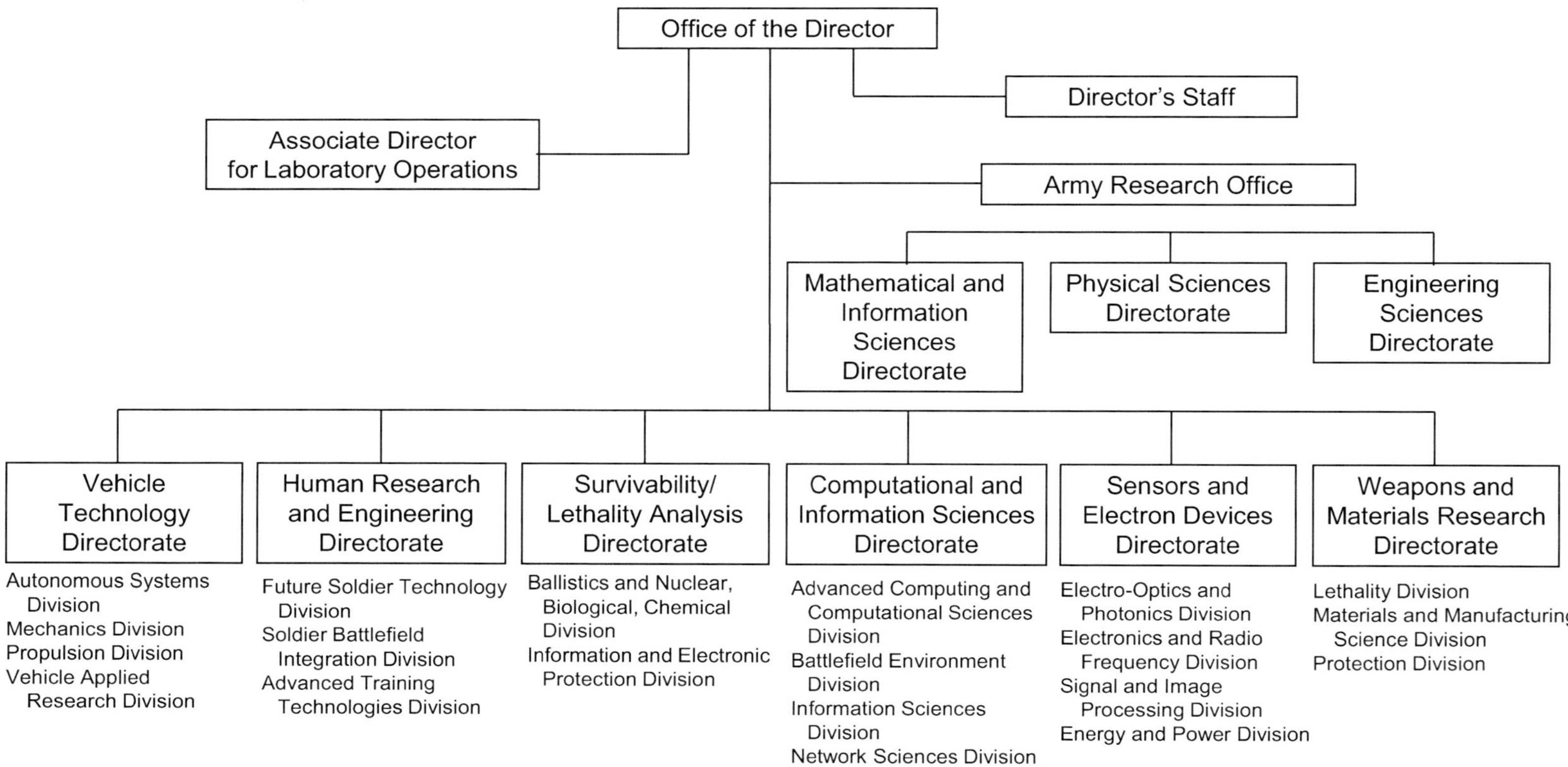

FIGURE A.1 Army Research Laboratory organization chart.

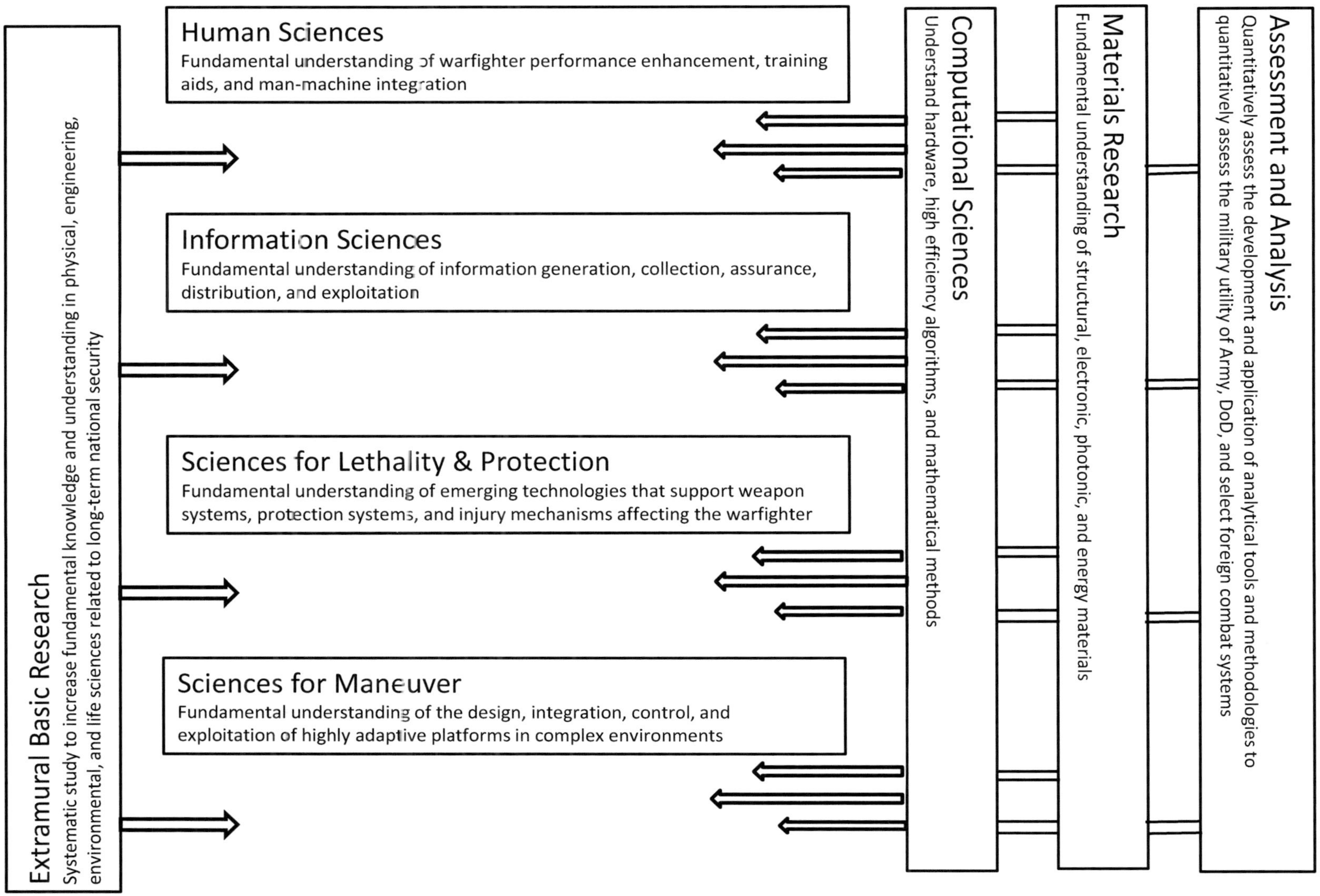

FIGURE A.2 The Army Research Laboratory's (ARL's) science and technology (S&T) campaign framework includes Extramural Basic Research, Computational Sciences, Materials Research, and Analysis and Assessment. These four crosscutting campaigns act in concert with its four focused campaigns, Human Sciences, Information Sciences, Sciences for Lethality and Protection, and Sciences for Maneuver, leading to knowledge products and technologies.

TABLE A.1 Mapping of the Army Research Laboratory (ARL) Organization Chart to the Science and Technology Campaign Areas Reviewed in 2015 and 2016

Campaign	Topic	ARL Directorate Involved
2015		
Materials Research	Biological and bioinspired materials Energy and power materials Engineered photonics materials	SEDD, WMRD
Sciences for Lethality and Protection	Battlefield injury mechanisms Directed energy Weapon-target interactions	WMRD, SLAD, SEDD, CISD
Information Sciences	Sensing and effecting System intelligence and intelligent systems	CISD, SEDD
Computational Sciences	Advanced computing architectures Computing sciences Data-intensive sciences Predictive simulation sciences	CISD, SLAD
Sciences for Maneuver	Human–Robot interaction Intelligence and control Perception	VTD, HRED, SEDD, CISD
Human Sciences	Humans in multi-agent systems Human variability Real-world behavior	HRED, SLAD
Analysis and Assessment	Assessing mission capability of systems	SLAD
2016		
Materials Research	High strain rate and ballistic materials Structural materials Electronic materials	SEDD, WMRD
Sciences for Lethality and Protection	Disruptive energetics and propulsion technologies Effects on target—ballistics and blast Flight, guidance, navigation, and control	WMRD, SLAD, SEDD, CISD
Information Sciences	Networks and communications Cybersecurity Human information interaction Atmospheric sciences	CISD, HRED, SEDD
Sciences for Maneuver	Platform mechanics Energy and propulsion Logistics and sustainability	VTD, SEDD
Human Sciences	Training Integration Technologies Augmentation	HRED, CISD, SLAD
Analysis and Assessment	Electronic warfare Cybersecurity Complex adaptive systems analysis	SLAD, SEDD, CISD, HRED

NOTE: CISD, Computational and Information Sciences Directorate; HRED, Human Research and Engineering Directorate; SEDD, Sensors and Electron Devices Directorate; SLAD, Survivability and Lethality Analysis Directorate; VTD, Vehicle Technology Directorate; and WMRD, Weapons and Materials Research Directorate.

B

Biographical Sketches of Army Research Laboratory Technical Assessment Board Members and Staff

JENNIE S. HWANG, *Chair,* is the chief executive officer of H-Technologies Group and board trustee and distinguished adjunct professor at Case Western Reserve. Her career encompasses corporate and entrepreneurial businesses, international collaboration, research management, technology transfer, and global leadership positions, as well as corporate and university governance. She has held senior executive positions with Lockheed Martin, SCM Corporation, and Sherwin Williams and has cofounded entrepreneurial businesses. She is internationally recognized as a pioneer and long-standing leader in the fast-moving infrastructure development of electronics miniaturization and green manufacturing. Dr. Hwang is an inventor and author of 350+ publications, including the sole authorship of several internationally used textbooks. As a columnist for the globally circulated trade magazines *Global Solar Technology* and *SMT* magazine, she addresses technology issues and global market thrusts. She is a member of the National Academy of Engineering and also has served on the International Advisory Board of the Singapore Advanced Technology and Manufacturing Institute and as a board director for Fortune 500 and private companies. Over the years, she has taught tens of thousands of professionals and managers in professional development courses, providing continuing education and disseminating new technologies to the workforce. The YWCA's Dr. Jennie S. Hwang Award was established to encourage and recognize outstanding women students in science and engineering. Her formal education includes the Harvard Business School Executive Program, a Ph.D. in materials science and engineering, two M.S. degrees, one in chemistry and one in liquid crystal science, and a bachelor's degree in chemistry.

KENNETH R. BOFF is principal scientist with Socio-Technical Sciences. From 2007 to 2012, he served as principal scientist with the Tennenbaum Institute at the Georgia Institute of Technology and as scientific advisor to the Asian Office of Aerospace Research and Development (Tokyo). From 1997 to 2007, he served as the U.S. Air Force Research Laboratory chief scientist for human effectiveness. In this position was responsible for the technical direction of a multidisciplinary R&D portfolio encompassing individual, organizational, and sociocultural behavior and modeling, training, protection, and the bio- and human engineering of complex systems. He is best known for his work on understanding and remediating problems in the transition of research to applications in the design, acquisition, and deployment of systems and the value-centered management of R&D organizations. Holder of a patent for rapid communication display technology, Dr. Boff has authored numerous articles, book chapters, and technical papers and is coeditor of *Organizational Simulation* (2005) and *System Design* (1987); he is also senior editor of the two-volume *Handbook of Perception and Human Performance* (1986) and the four-volume *Engineering Data Compendium: Human Perception and Performance* (1988). He actively consults and provides technical liaison with government agencies, international working groups, universities, and professional societies. He has organized and facilitated numerous technical workshops in the United States and Europe and along the Pacific Rim focused on contemporary issues in complex sociotechnical systems. He is a fellow of the Human Factors and Ergonomics Society and the International Ergonomics Association.

MARK EBERHART is a professor in the Department of Chemistry and Geochemistry at the Colorado School of Mines, where he directs the Molecular Theory Group (MTG). At the MTG, knowledge of bonding is obtained through detailed topological analyses of the spatial distribution of electrons in molecules and solids. Many subtle aspects of the distribution become obvious when viewed from a topological perspective. The accompanying topological formalism gives well-defined, unambiguous, meaningful, and consistent definitions to previously indeterminate quantities such as atomic bonds and basins. His work is based primarily on first principles computations, which provide the electron charge densities, and on topological analysis software developed at the MTG. He is also exploring the topological and geometric origins of the stability of amorphous metallic alloys. In addition to its work on condensed-phase systems, his group has active research programs exploring the relationships between charge density and the chemical properties of molecular systems, both organic and inorganic. Dr. Eberhart holds a B.S. in chemistry and applied mathematics from the University of Colorado, an M.S. in physical biochemistry from the University of Colorado, and a Ph.D. in materials science and engineering from the Massachusetts Institute of Technology.

GEORGE T. (Rusty) GRAY III is a laboratory fellow and staff member in the dynamic properties and constitutive modeling team within the Materials Science Division of Los Alamos National Laboratory (LANL). He came to LANL following a 3-year visiting scholar position at the Technical University of Hamburg-Harburg in Hamburg, Germany, having received his Ph.D. in materials science in 1981 from Carnegie Mellon University. As a staff member (1985-1987) and later team leader (1987-2003) in the Dynamic Materials Properties and Constitutive Modeling Section within the Structure/Property Relations Group (MST-8) at LANL, he has directed a research team working on investigations of the dynamic response of materials. He conducts fundamental, applied, and focused programmatic research on materials and structures, in particular in response to high-strain-rate and shock deformation. His research is focused on experimental and modeling studies of substructure evolution and mechanical response of materials. These constitutive and damage models are utilized in engineering computer codes to support large-scale finite element modeling simulations of structures ranging from national defense (DOE, DOD, DARPA), industry (GM, Ford, Chrysler, and Bettis), foreign object damage, and manufacturing. He is a Life Member of Clare Hall, University of Cambridge, in the U.K., where he was on sabbatical in the summer of 1998. He co-chaired the Physical Metallurgy Gordon Conference in 2000. He is a fellow of the American Physical Society (APS), a fellow of ASM International (ASM), and a fellow of the Minerals, Metals, and Materials Society (TMS). He also serves on the International

Scientific Advisory Board of the European DYMAT Association. In 2010 he served as the president of the Minerals, Metals, and Materials Society. Starting in 2012 he became the chair of the Acta Materialia board of governors, which oversees the publication of the journals *Acta Materialia*, *Scripta Materialia*, and *Acta Biomaterialia*. He has authored or coauthored over 410 technical publications.

PRABHAT HAJELA is provost and professor of mechanical and aerospace engineering at the Rensselaer Polytechnic Institute. His research interests include analysis and design optimization of multidisciplinary systems; system reliability; emergent computing paradigms for design; artificial intelligence; and machine learning in multidisciplinary analysis and design. Before joining Rensselaer, he worked as a research fellow at the University of California, Los Angeles, for a year and was on the faculty at the University of Florida for 7 years. He has conducted research at NASA's Langley and Glenn Research Centers and the Eglin Air Force Armament Laboratory. In 2003, Dr. Hajela served as a congressional fellow responsible for science and technology policy in the Office of U.S. Senator Conrad Burns (R-Mont.). He worked on several legislative issues related to aerospace and telecommunications policy, including the anti-SPAM legislation that was signed into law in December 2003. Dr. Hajela is a fellow of the American Institute of Aeronautics and Astronautics (AIAA), a fellow of the Aeronautical Society of India (AeSI), and a fellow of the ASME. He has held many editorial assignments, including editorship of *Evolutionary Optimization* and associate editorship of the *AIAA Journal*, and is on the editorial boards of six other international journals. He has published over 270 papers and articles in the areas of structural and multidisciplinary optimization and is an author or coauthor of four books in these areas. In 2004, he was the recipient of AIAA's Biennial Multidisciplinary Design Optimization Award.

WESLEY L. HARRIS is the Charles Stark Draper Professor of Aeronautics and Astronautics, Associate Provost for Faculty Equity, and Director of the Lean Sustainment Initiative at the Massachusetts Institute of Technology. He was elected to the National Academy of Engineering "for contributions to understanding of helicopter rotor noise, for encouragement of minorities in engineering, and for service to the aeronautical industry." He has performed research and published in refereed journals in the following areas: fluid mechanics; aerodynamics; unsteady, nonlinear aerodynamics; acoustics; lean manufacturing processes; and military logistics and sustainment. Dr. Harris has substantial experience as a leader in higher education administration and management. He also has demonstrated outstanding leadership in managing major national and international aeronautical and aviation programs and personnel in the executive branch of the federal government. He is an elected fellow of the AIAA, the AHS, and the NTA for personal engineering achievements, engineering education, management, and advancing cultural diversity.

Staff

LIZA HAMILTON is an associate program officer at the Laboratory Assessments Board (LAB) in the Division on Engineering and Physical Sciences (DEPS) at the National Academies of Sciences, Engineering, and Medicine. Since 2002, she has been responsible for managing the administrative aspects of panel formation, panel meetings, report publication and dissemination, and program development. In addition, she has designed newsletters, brochures, covers, and figures for numerous reports prepared by the Division on Life Sciences and DEPS of the National Academies of Sciences, Engineering, and Medicine. Ms. Hamilton earned a 4-year certification in musical theater performance from Pinellas County Center for the Arts in St. Petersburg, Florida; a B.F.A. in film studies from the University of Utah; a design certification from Maryland Institute College of Art; and the master of liberal arts from the Johns Hopkins University.

EVA LABRE is the administrative coordinator for the LAB. Since 2009, she has been responsible for assisting in the management of the administrative aspects of panel formation, panel meetings, report publication and dissemination, and program development. In addition, she has been responsible for travel expense accounting. In February 2014, she was promoted and has recently taken on more responsibilities related to financial aspects of the work of the LAB. Ms. Labre previously held administrative positions at the National Academies on the staff of the Committee on International Organizations and Programs in the Office of International Affairs and on the staff of the Research Associateship Program in the Office of Scientific and Engineering Personnel. Ms. Labre has a B.A. in art history from George Washington University.

JAMES P. McGEE is the director of the LAB, the Army Research Laboratory Technical Assessment Board (ARLTAB), and the Committee on the National Institute of Standards and Technology Technical Programs, all within DEPS at the National Academies. Since 1994, he has been a senior staff officer at the Academies, directing projects in the areas of systems engineering and applied psychology, including activities of ARLTAB and projects of the Committee on National Statistics' Panel on Operational Testing and Evaluation of the Stryker Vehicle and the Committee on Assessing the National Science Foundation's Scientists and Engineers Statistical Data System, the Committee on the Health and Safety Needs of Older Workers, and the Steering Committee on Differential Susceptibility of Older Persons to Environmental Hazards. He has also served as staff officer for the National Academies' projects on air traffic control automation, musculoskeletal disorders and the workplace, and the changing nature of work. Prior to joining the Academies, Dr. McGee held technical and management positions in systems engineering and applied psychology at IBM, General Electric, RCA, General Dynamics, and United Technologies. He received his B.A. from Princeton University and his Ph.D. from Fordham University, both in psychology, and for several years instructed postsecondary courses in applied psychology and in organizational management.

ARUL MOZHI is senior program officer at the LAB. Since 1999, he has been directing projects in the areas of defense science and technology, including those carried out by numerous study committees of the LAB, the ARLTAB, the Naval Studies Board, and the National Materials and Manufacturing Board. Prior to joining the National Academies, Dr. Mozhi held technical and management positions in systems engineering and applied materials research and development at UTRON, Inc.; Roy F. Weston, Inc.; and Marko Materials, Inc. He received his M.S. and Ph.D. degrees (the latter in 1986) in materials engineering from the Ohio State University and then served as a postdoctoral research associate there. He received his B. Tech. in metallurgical engineering from the Indian Institute of Technology, Kanpur, in 1982.

C

Assessment Criteria

The Army Research Laboratory Technical Assessment Board's (ARLTAB's) assessment considered the following general questions posed by the ARL director:

- Is the scientific quality of the research of comparable technical quality to that executed in leading federal, university, and/or industrial laboratories both nationally and internationally?
- Does the research program reflect a broad understanding of the underlying science and research conducted elsewhere?
- Does the research employ the appropriate laboratory equipment and/or numerical models?
- Are the qualifications or the research team compatible with the research challenge?
- Are the facilities and laboratory equipment state of the art?
- Are programs crafted to employ the appropriate mix of theory, computation, and experimentation?

To assist ARL in addressing promising technical approaches, the Board will also consider the following questions:

- Are there especially promising projects that, with improved direction or resources, could produce outstanding results that can be transitioned ultimately to the field?
- Are there promising outside-the-box concepts that should be pursued but are not currently in the ARL portfolio?

The ARLTAB applied the following metrics or criteria to the assessment of the scientific and technical work reviewed at the ARL:

Project Goals and Plans

- Are the objectives clearly stated and are tasks well defined to achieve objectives?
- Are milestones defined? Are they appropriate? Do they appear feasible?
- Are obstacles and challenges defined (technical, resources, time)?
- Assuming success, what difference will it make to the science base, the end-user, or in a mission area context?
- Does the project plan identify dependencies (i.e., successes depend on success of other activities within the project or on the success of projects developed outside ARL)?
- Does the project represent an area where application of ARL strengths is appropriate?
- What stopping rules, if any, are being or should be applied?

Methodology and Approach

- Are the methods (e.g., laboratory experiment, modeling/simulation, field test, analysis) appropriate to the problems? Do these methods integrate?
- Are the hypotheses appropriately framed within the literature and theoretical context?
- Is there an alternative approach that facilitates the progress of the project?
- Is there a clearly identified and appropriate process for performing required analyses, prototypes, models, simulations, tests, etc.?
- Is the data collection and analysis methodology appropriate?
- Are conclusions supported by the results?
- Are proposed ideas for further study reasonable?
- Do the trade-offs between risk and potential gain appear reasonable?
- If the project demands technological or technical innovation, is that occurring?

Capabilities and Resources

- If staff or equipment are not adequate, how might the project be triaged (which technical thrust should be emphasized, which sacrificed?) to best move toward its stated objectives?
- Recruiting new talent into ARL.

Scientific Community

- Presentations and colloquia.
- Participation in professional activities (society officers, conference committees, journal editors).
- Papers in quality refereed journals and conference proceedings (and their citation index).
- Educational outreach (serving on graduate committees, teaching/lecturing, invited talks, mentoring students).
- Fellowships and awards (external and internal).
- Participation on review panels (ARO, NSF, MURI, . . .).
- Patents and intellectual property (IP) and examples of how the patent or IP is used.
- Involvement in building an ARL-wide cross-directorate community.
- Public recognition, e.g., in the press and elsewhere, for ARL research.
- Collaborations (lead, partner, support).

D

Acronyms

2D	two-dimensional
3D	three-dimensional
ABD	avalanche breakdown diode
ABLE	atmospheric boundary layer environment
ACM	Association for Computing Machinery
AESA	Active Electronically Steered Array
AFDD	Army Aero-Flight-Dynamics Directorate
AFM	atomic force microscopy
AFRL	Air Force Research Laboratory
ANL	Argonne National Laboratory
ARDEC	U.S. Army Armament Research, Development and Engineering Center
ARL	Army Research Laboratory
ARLTAB	Army Research Laboratory Technical Assessment Board
ARO	Army Research Office
ARW	advanced research WRF (weather research and forecasting)
CASA	complex adaptive systems analysis
CCE	core campaign enabler
CFD	computational fluid dynamics
CISD	Computational and Information Sciences Directorate
COTS	commercial off-the-shelf
CRA	Collaborative Research Alliance
CRADA	cooperative research and development agreement

CS	compressive sensing
CSMA/CA	carrier sense multiple access with collision avoidance
CSR	Center for Systems Reliability
CT	computerized tomography
CTA	Collaborative Technology Alliance
DARPA	Defense Advanced Research Projects Agency
DE	directed energy
DFT	density functional theory
DIC	digital image correlation
DMECS	deployable multi-sensor electromagnetic warfare characterization system
DOD	Department of Defense
DOE	Department of Energy
DPD	dissipative particle dynamics
DRFM	digital radio frequency memory
DU	depleted uranium
EAR	environment for auditory research
EEG	electroencephalogram
EME	electromagnetic environment
ERE	entities, relations and event
EW	electronic warfare
FEM	finite element method
fMRI	functional magnetic resonance imaging
FPGA	field programmable gate array
FY	fiscal year
GIFT	Generalized Instructional Framework for Tutoring
GN&C	guidance, navigation and control
GPS	Global Positioning System
GPU	graphics processing unit
GUI	graphical user interface
HII	human and information interaction
HPC	high-performance computing
HRED	Human Research and Engineering Directorate
HRI	human–robot interaction
HSI	human–system integration
HTML	HyperText Markup Language
IC	integrated circuit
ICT	Institute for Creative Technology
IED	improvised explosive device
IEEE	Institute of Electrical and Electronics Engineers
IMPRINT	Improved Performance Research Integration Tool

IR	infrared
IT	information technology
ITA	International Technology Alliance
JP-8	jet propellant 8 (fuel)
KCI	key campaign initiative
LAPS	local analysis and prediction system
LBM	Lattice Boltzmann Method
LES	large-eddy simulation
Li-ion	lithium ion
M&S	modeling and simulation
MAC	Medium Access Control
MAST	micro-autonomous systems and technology
MASTODON	Mountainous Slope Transport and Diffusion
MATERHORN	Mountain Terrain Atmospheric Modeling and Observations
MD	molecular dynamics
MEMS	microelectromechanical system
MIMO	Multiple Input, Multiple Output
MODE	method for object-based diagnostic evaluation
MOSFET	metal-oxide semiconductor field-effect transistor
MOVES	Modeling, Virtual Environments and Simulation
MSA	meteorological sensor array
MURI	Multidisciplinary University Research Initiative
NASA	National Aeronautics and Space Administration
NC	networks and communications
NIE	Network Integration Evaluation
NSA	National Security Agency
NSF	National Science Foundation
NSRL	Network Science Research Laboratory
NWP	numerical weather prediction
OMEN	Optimized Modular EW Network
OneSAF	One Semi-Automated Force
OTA	Other Transaction Authority
PDV	photon Doppler velocimetry
PE	polyethylene
PI	principal investigator
PLDI	programming language design and implementation
QMU	quantification of the margin of uncertainty
QWIP	quantum-well infrared photodetector

R&D	research and development
RANS	Reynolds Averaged Navier-Stokes
RAS	remote and autonomous system
RBD	rigid-body-dynamics
RCAS	Rotorcraft Comprehensive Analysis System
RDEC	Research, Development and Engineering Center
RF	radio frequency
RIP	Routing Information Protocol
ROMP	ring-opening metathesis polymer-based composite
S&T	science and technology
SEDD	Sensors and Electron Devices Directorate
SLAD	Survivability/Lethality Analysis Directorate
SMART	smart material actuated rotor technology
SNL	Sandia National Laboratories
SPEAR	soldier performance and equipment advanced research
TARDEC	Tank Automotive Research, Development, and Engineering Center
TBI	traumatic brain injury
TDMA	time division multiple access
TEM	transmission electron microscopy
TIREM	Terrain Integrated Rough Earth Model
TN	translational neuroscience
TRU	Technology Refresher Update
TTCP	Technical Cooperation Program
TWT	traveling-wave tube
UAS	unmanned aircraft system
UHMWPE	ultrahigh molecular weight polyethylene
UV	ultraviolet
V&V	validation and verification
VI	vehicle intelligence
VRAMS	virtual risk-informed agile maneuver sustainment
VT	vehicle technology
VTD	Vehicle Technology Directorate
WC	tungsten carbide
WMRD	Weapons and Materials Research Directorate
WRE-N	weather running estimate—nowcast
WRF	weather research and forecasting
WSMR	White Sands Missile Range